The Microbial State

The Microbial State

Global Thriving and the Body Politic

Stefanie R. Fishel

 University of Minnesota Press
Minneapolis
London

Portions of the Introduction, chapter 1, and chapter 2 were previously published as "Microbes," in *Making Things International 1: Circuits and Motion*, ed. Mark B. Salter (Minneapolis: University of Minnesota Press, 2015).

Copyright 2017 by the Regents of the University of Minnesota

All rights reserved. No part of this publication may be reproduced, stored in a retrieval system, or transmitted, in any form or by any means, electronic, mechanical, photocopying, recording, or otherwise, without the prior written permission of the publisher.

Published by the University of Minnesota Press
111 Third Avenue South, Suite 290
Minneapolis, MN 55401-2520
http://www.upress.umn.edu

A Cataloging-in-Publication record for this book is available from the Library of Congress.

ISBN 978-1-5179-0012-0 (hc)
ISBN 978-1-5179-0013-7 (pb)

The University of Minnesota is an equal-opportunity educator and employer.

To all manner of fantastic creatures that brought this book life, love being the most mythical and wonderful of all.

Contents

Preface and Acknowledgments	ix
Introduction: Involutionary Politics	1
1. Corporeal Politics	25
2. Lively Subjects, Bodies Politic	55
3. States in Nature, Nature in States	76
4. Posthuman Politics	98
Coda: New Metaphors for Global Living	113
Notes	119
Index	141

Preface and Acknowledgments

This book started with a bad case of pneumonia. This is an unusual beginning to be sure, but one that prompted an extensive reappraisal of my academic discipline—International Relations (IR)—and my experiences during graduate school at Johns Hopkins University in Baltimore, Maryland. These reconsiderations of the personal and professional became the motivation to think about my research in new ways. This led to refiguring the way in which the state, and the bodies within it, is understood in theory and practice.

While teaching and taking classes during my doctoral studies, I also came to suffer the enduring effects of a weakened and stressed immune system from misdiagnosed pneumonia and misuse of antibiotics. My body's troubles turned my thoughts toward how we understand and respond to the world around us based on our bodily experience of it. Moreover, as a theorist with materialist leanings, I was challenged to think beyond the mainstream paradigms of IR based on realist assumptions of the nastiness and brutishness of life, or the utopic desires of liberalism, because of my day-to-day encounters on my street and in the city that had become my home. The two-dimensional "rational, self-interested actor" did not provide much insight into the complex interactions, desires, and actions both generous and egoistic that I witnessed. The world, with its cruelty and generosity, was more robust and held more hope than this anemic theory could convey with its belief in anarchic structures, self-interested subjects, and its metaphors of "survival of the fittest" and "balances of power." As my research progressed, so too did my conviction that IR needed to explore new words and concepts to reflect the "real" world we experience everyday.

Then, as I mentioned above, my pneumonia went viral and uncovered a web of food allergies due to a compromised bacterial community harmed by excessive antibiotics prescribed to treat a viral infection. This led me ever deeper into the theories of the body and new studies about its constitution.

The gut microbiome and the human immune system revealed themselves to be rich subjects to explore IR's notions of the body politic, or the metaphorical story of how the body comes to explain and naturalize political relations. My body became a source of insight and an object of study. I turned these experiences and insights toward International Relations.

It became clear after preliminary research that IR did not pay much attention to bodies (and even less to bacteria), but rather to individuals, citizens, or persons; these are legal or economic understandings of human existence and can sometimes be attached to corporeal existence, but are most often defined by law or social conventions. This is not to say these do not matter, but I felt the fleshiness of the body itself should be reckoned with in more detail and with more care. As a theorist, this bodily lack provoked me, for much of IR scholarship also had not thought critically about its reliance on the concept of the "body politic." IR did not talk much about bodies, but it still relied on the body politic as a guiding metaphor for understanding states. Where had the body gone? What body did it assume sat next to the "politic"? As I will stress in this book, much has changed since Hobbes, Rousseau, and Locke wrote on sovereignty and the body politic. Gender and race scholars have called attention to how the form of the body matters in all things, and I hope this book expands and enriches this theoretical emphasis.

This is all to say that this book was born crowded and lively. My body and life contained multitudes, to paraphrase Whitman, as I shall do throughout, and this was more often than not an uneasy mosaic of shifting alliances and misunderstood impulses. Most of the time we lived together gracefully, but at crucial moments I was reminded that, just as the Westphalian sovereign state is questioned by the realities of an interconnected world, my body and state could no longer be imagined as a container. It was a place of diverse and often mysterious "actors" on a shifting stage.

I desire, with this book, to imagine political structures that not only support many forms of life but also discover *new* ones that celebrate life in all its varied, magical forms to acknowledge the motley crew of influences at all levels of life and politics. I do not think we should aim for mere living: we need to set our sights on building a world for thriving. What might a life worth living be for multiple species, not just the human animal? My argument, in brief, is that this can begin by merely seeing and talking about the world around us differently. Novel metaphors and analogies can restore enchantment and complexity to political debates that have found themselves stymied and confounded by language that no longer reflects the global flows around us. Bodies, with all their messy complexities and relationships with other bodies, can still be used as models for the body politic—not a Leviathan, but a lively vessel of coevolved partners and messmates.

I write this book as an intervention into International Relations; not another in the familiar debates about whether or not the international community is justified in using military action, but one that hopes for future scholars to find bodies, in all their complexity, present in IR along with the body as an object of study in its own right; not limited by the state and its security as a primary emphasis of the discipline.

To this end, the book is divided into three broad sections. The Introduction and first chapter build a framework that places bodies and their corporeal forms at the center of IR. It does this through theories that inspire and sustain material conceptions of politics and through the study of metaphor as ground in the experiences of our bodies in space. The second part explores two illustrative examples from the biological sciences where the interior of the human body and its processes guide IR into new metaphors for both bodies and the states in which they find themselves. Importantly, these examples from medicine and biology help explain what is happening in the world more effectively, and affectively, than other approaches in mainstream IR. The final chapter offers a reimagining of the politics of life that can look to future political structures with positivity and reality, in its most real and physical sense.

Along with my microbial messmates, both pathogenic and commensal, this book is joined by many other influences on my thinking and writing; it is an assemblage and a multiplicity. I am indebted to many individuals who helped and inspired me during the research and writing of this book. My circle of friends, mentors, family, and loves only make my life and work as an academic richer. My professors and colleagues at University of Victoria and Johns Hopkins: Rob Walker for early direction on my master's thesis; Siba Grovogui and Jane Bennett for their careful and patient supervising of the research that contributed to this book; Bill Connolly, Dick Flathman, and Renee Marlin-Bennett for their kind and generous scholarly souls; my friends and fellow soldiers-in-arms during graduate school and beyond with special thanks to Anthony Burke, Daniel Levine, Jairus Grove, Ben Meiches, Lauren Wilcox, Nico Taylor, and Jim Morrow for their encouragement and practical help.

I would be remiss to leave out the special contribution of my family: Griffin, the one who makes me want to fight for this fragile world we live in together. Marc, who has traveled each step of this journey with me. Kathleen Linden, a mother who raised strong and independent daughters, and Sydney Linden, a woman who fights every day for a more just world. And lastly, Gertrude, who was unable to see this to the end, but who remains my fierce soul sister and messmate.

I benefited from the support of a writing grant from the AAUW in 2011. This organization for women in the university funded a year of writing for this project. The University of Alabama's gender and race department is the home where I was able to make this book a reality.

To the reader, take this book in the spirit in which it was written: the material in it, and the author, have been deeply affected by a wider interest in creating a world that is just and ecologically sustainable for all creatures. I am glad you are here with me.

Introduction

Involutionary Politics

I left no one at the door, I invited all;
The thief, the parasite, the mistress—these above all I called—

—Walt Whitman, *Leaves of Grass*

International Relations needs a bigger vocabulary. This claim does not mean that we need a more specialized language or theoretical jargon, but rather new words and concepts that explain the world with greater clarity. It means, as in the epigraph by Whitman above, we open our door to those who have been excluded or ignored at both a disciplinary level and a worldly one. We can invite guests from other disciplines or redraw the intellectual history of International Relations (IR) and reuse it for a new era of global or, more hopefully, planetary politics. This could begin simply with giving up the title "International Relations." This discipline and the world it explains are more than, and less than, relations between nations. The familiar IR view of states and their corresponding nations obfuscates the challenges facing human communities in what has become an epoch named after human alterations to our planetary ecosystems, dubbed the Anthropocene.[1]

Responding to challenges in this epoch—sea level rise, increased carbon dioxide emissions, deforestation, maldevelopment, and mass extinction of nonhuman species—alongside the corresponding human and nonhuman misery these will bring, must be incorporated into all levels of policy, activism, and intellectual activity.[2] The planet does not need saving, but the conditions that support all life on earth must be preserved and nurtured. It is this shift in concern and scale that deserves full attention and care.

Within this larger planetary crisis, it is difficult to speak of the Anthropocene and the changes to our world from specific disciplines or economic theories. The planet demands more from us than the unexamined ideas of human mastery over nature that shut out other kinds of relations with both ourselves and other beings and that are found in most approaches to the international or global. This obsession with mastery leads to seeing the planet as a stage where a grand story unfolds with humans performing all the lead roles, or as merely a resource that is exploited for monetary profit, personal or national gain. Social theory and the natural sciences, even at their most progressive, often aid in supporting these narrow anthropocentric and instrumentalist understandings of relations only within and across human communities, rather than across ecologies, embedded in the environment and life-giving systems, and existing within the planetary biosphere.

To battle these tendencies, scholars, and perhaps those who could be called a global citizenry more broadly, must be able to shift perceptions between the small and the large, from microbes to the biosphere, and be able to switch lenses from the view of the human to that of the nonhuman. It is important to recognize that it matters what figures we think through, from, and with.[3] If this thinking always comes from "man," and particular social understandings of human bodies, then it is the goal of any response to highlight the assumptions that have heretofore been taken as truth about this body, about "man." *The Microbial State* uses the body, the body politic, and its attendant metaphors to bring specificity to these broader disciplinary and political concerns.

In other words, it is not that man, anthropos, or the body itself is undertheorized or well considered, but rather that it has been assumed to be a particular kind of body—"man" is the not the neutral designation it is portrayed to be. As feminist, postcolonial, and poststructural approaches stress, man is historically constructed through the experience of white European males, and through power and violence, assuming a universality that has silenced and created a "unified, consistent identity" that does not reflect the dispersed and hybrid subjectivity as seen by postcolonial and poststructural theorists.[4] As Sylvia Wynter writes, our present conception of man "overrepresents itself as if it were the human itself."[5] In addition, treating the body as if it has no specificity is also harmful; to ignore race and racism only serves to make race appear as a biological given where one specific group (white European male) embodies humanity.[6] This places IR scholars in the sometimes unrecognized and other times deliberate support of the legacy of white supremacy in the global order, its institutions, and IR as the discipline that claims to speak for it.[7]

The Microbial State challenges the reader to think of politics as immanent to the biospheric rather than the anthropic: we humans live in many overlapping worlds and experience many globalizations depending on where we find ourselves geographically. These politics are involute, composite, often discordant, but entangled within and supported by the biosphere we share; a biosphere that is quickly becoming unable to support much of the life we see on it today. We will have to open our hearts to matters of concern that are multidisciplinary, multiracial, and multigendered. We must be sensitive to oppression and the needs of the most vulnerable, to be open to viewpoints from multiple origins, and to think creatively across species lines. The future demands poets and visionaries.

Ethical Interventions

This future focus urges the reader to reflect on the conditions of living globally, and on how humans change, both individually and socially, with a desire to affirm and celebrate life. This is necessarily a delicate ethical engagement in a world where harms are shared across worlds and temporal spaces. The desire for change leads me to questions of design, both linguistic and material. New metaphors appear to shape the rhetoric and thinking about how the human capacity to create might include nonhuman others. I begin with language and metaphor because all creation includes abstractions, like language, as an integral part of making. "In other words, mental skills and thought processes—the ability to use 'mind tools,' which represent and articulate concepts of what might be—are as essential in any productive process" as any tool used by hand.[8] To ponder the imbrication of word and world begins with corporeal metaphors and the body politic and then extends these through reflections on how new political spaces may be conceived and built.

With these preliminary reflections in mind, we should ask ourselves, as scholars and global citizens, questions about how we might imagine and design global politics. If we step back from the practical concerns of how one creates new political spaces for a global demos, or the hubris that may have to be admitted before one could even begin to describe what may be needed in response to the crises and radical possibilities that humans face in the twenty-first century, where does one start? It is vital that these spaces should not always be solutions to previously defined problems, or focused on the seemingly given objects or concepts that embody or seem to embody these problems. Rather, we could extend this idea of designing a world much further into a "dynamic process of innovation and refinement beyond the constraints of time and space."[9] For IR, this means questioning the state, as it is the dominant mode of understanding politics through time and space, temporality and territory.[10]

Moreover, after implementing new designs based on these refigured spaces, there must be a continued process of evaluation and redesign, if necessary. If we think of ourselves as designers of these worlds with the recognition that these worlds do not always belong to humans, who instead coinhabit them with others. The very practice of design is about making choices, and these choices imply responsibility and an ethical consideration of how one design is chosen over another. This could also mean adapting former techniques and forms to fit new conditions. Democracy, by way of example, may need as much attention as alternative energy design to make it more responsive to the changing needs of people in the twenty-first century. Freedom may mean less about individuality and more about finding connections between diverse actors.

This begins with being frank about where we are starting from and what small steps might be taken to shift the conversation in IR from state power and national primacy to care for bodies, both human and nonhuman, in an environment that is unpredictable and often violent. This is especially crucial because, more often than not, states are perpetrators or supporters of violence against its citizens and others. This may mean admitting a certain amount of disciplinary failure in International Relations' continued support of the state as an object of study. I believe this means welcoming diverse disciplinary guests to the table, asking for advice, and being honest about where we have gone wrong and why.

This chapter introduces the guests invited to build a framework of thought for later chapters. States, the role of science, the theoretical approaches used in the book, and a preliminary explication of what ethics may follow will be introduced and defined. I offer a framework for thinking about shared spaces in which we can live, and even thrive, together. This builds the backdrop needed to engage with later chapters on bodies and words, bacteria and biomes, immunity and community, and finally life and the future.

My tools of choice for expanding perspectives in IR are science and technology studies and the biological sciences, paired with the familiar metaphorical constructions of the body politic and the state imagined as a person. Together, these approaches lend conceptual innovation, techniques for studying the relations between and within humans and nonhumans, and practical methodology to the study of IR; metaphor builds a bridge between language and materiality, allowing for new and much needed conversations to come forward in IR. This does not mean a purely immunological, essentialist, or biological model of the body is endorsed, but instead that this combination can offer ways to understand complex global interactions in a new light. In a world of incomplete or broken models, another model will not save us, but framing bodies and states with more complexity may suggest new configurations for political and social organization.[11]

Microbial Interventions

To find new configurations, I take microbes as an object of study for demonstrating that the relationship between humans and microbial communities give a bodily and material example in which multiple perspectives and objects—beyond human-created institutions and subjects—can be seen as vital and necessary to politics and human survival. This is both a metaphorical and actual relationship. Humans survive with the help of microbial communities, and these ties can be used analogically to better understand human institutions, politics, and community creation. The addition of the microbes to these relations offers a view of a complex world of interactions between human bodies and their microbiomes. They unsettle the human's place in the great chain of being and allow the agentic capability of nonhuman actors to be noted and appreciated. This adds materiality and agency to the body (both human and nonhuman) to supplement the discursive focus on the body as text, information, or social construction, and adds "lively life" to the bleak theological and transcendent analyses of the body as "bare life" or zoë.[12]

While keeping these important considerations in the forefront, the human body, as it is understood through its relation with microbes and parasites, will work together with theoretical approaches to demonstrate that giving the body specificity within the discipline can lend itself to analogizing how political bodies like the state can be better understood. Based on these novel scientific and social understandings of human bodies, and their capacities and interaction in the world, social theory is dislocated from the epistemologies it relies upon. These dislocations and capacity building can transform and enable human subjects and political systems by demonstrating that embodiment, and the process of creating a subject, is necessary to facing new political and social challenges.

It is of paramount importance to reach beyond an andro- and anthropocentric focus in IR to encourage an ethic of care and an ethos of love for the earth beyond what value its resources and its creatures can offer the human animal and its corporations and states. Human impact on the earth is profound and demands global political action to counter and ameliorate the effects of human terraforming and development.

In this way, microbes are not merely a theoretical construct used to create discussion, but rather a way to think past anthropocentrism. This provocation to theorize comes from microbes themselves, rather than stemming from a search to find a "trick" or subject matter that illuminates my theoretical commitments. Microbes as an object of study demonstrate that the relationship between humans and microbial communities gives a bodily and material example in which multiple perspectives and objects—beyond human-created institutions and subjects—can be rightly understood as vital and necessary to politics and human survival.

Putting aside that microbes, made up of single-celled organisms and viruses, are the most common living thing on earth, I love the idea that we are walking assemblages with our own microbiotic communities (as distinct as a fingerprint) and that globally, closely associated communities of microbes are key players in everything from photosynthesis to human digestion.[13] In an interview, Donna Haraway said that she was irreducibly "in love with real critters" and that her work "tries to take seriously the fact that all love objects are inappropriate. If you are actually in love, you find yourself always to be in love with the wrong kind of love object . . . love undoes and re-does you. And this love can land you in interesting places and re-figure all kind of relationships."[14] This love calls one to be a collaborationist at heart and be "struck," as Jane Bennett writes in *Vibrant Matter*, by the lively presence of the countless billions of microbes that are essential for life on earth.[15] This approach helps to foster a picture of the global order that does not just think about human bodies in situ, but instead about bodies in motion and the relationships that link bodies together. If we begin to think about the world through these connections and linkages, personal autonomy through plural distinctiveness, we come closer to a politics that acknowledges shared forms of life. We need to reform and re-figure our kin networks.[16]

Posthuman Interventions

The book will guide these ideas of politics in a broadly posthumanist direction to counteract the humanist, or anthropocentric, tendencies in the social sciences. A posthuman intervention urges cross-disciplinary thinking, especially between the social and natural sciences. The social sciences tend to focus on the human and treat the social world as if it is a separate reality from the "real" world—a place of subjective knowledge. When contrasted with the sciences, a range of disciplines that study a world where people are absent and objectivity reigns, we discover a dualism within our disciplines themselves: a world of people and a world of things kept separate through method and ideology. Methodologically, following Andrew Pickering's formulation, this demands that the researcher find a perspective where the human (people) and nonhuman (things) are not separate matters of inquiry; rather one sees with double vision.[17] This means disregarding disciplinary divisions between the hard sciences as studying only the world of things, or the real, and the soft sciences as in control of the world of people, or the social. This shift, this seeing double, can be defined as a shift to a posthuman perspective. This double vision creates a third object of inquiry: one can see a wider range of agency and action. "We immediately get the idea of a *mutual becoming* of coproduction or of coevolution of people and things. This mutual becoming is precisely what the posthumanist shift brings into focus" (31). In other words, looking at the human, as either a social or biologically constructed unit, separately from the nonhuman or the environment it exists within, cannot offer a full view of how these bodies work in common. Pickering writes, "this would miss something about being in the world . . . the reciprocal production of subjects and objects, the coupled becoming of the human and nonhuman" (34).

It is imperative to mix and overlap the previously prescribed concepts that traditional approaches offer in an attempt to open up new spaces of academic inquiry. This may mean eschewing disciplinary boundaries and even encouraging new disciplines to form based on the problems that need to be addressed. Traditional approaches often do not capture social and material environments in their full complexity and enmeshments. Beyond a methodological commitment to multidisciplinarity in conceptual analysis, the approach this book creates through the body, metaphor, and a posthuman perspective is also a challenge to think differently about the composition of the world. The focus on the figure of the posthuman in the fourth chapter is not a facile temporal argument about what may come after the human in an evolutionary or biological sense, but rather one that can redefine political ontology, agency, and efficacy at the global level. The posthuman is shorthand for thinking together across disciplines, and through bodily and worldly experiences.

By way of example, this might mean reconsidering the discourse of biology and its reliance on species differentiation. What if biologically partitioning homo sapiens as species is not just a scientific act, but a political one that reinforces hierarchy, abandonment, and appropriation of all other species found on the planet? That in fact, according to Ed Cohen, "contestations between and among different 'natures' constitute the actual matter of politics."[18] As Cohen further questions: "How did we come to believe that 'belonging to the human species' is the best way to make sense of 'being human' in the first place?"[19] To be human as species automatically erases and makes abject those that are not human, those that are Other. To place this in a social context, historians and social theorists know the harm that social Darwinism, as an understanding of homo sapiens broken across specious racial lines, did when twinned with a rapacious need for raw materials to feed the industrial revolution: it became a justification for slavery, murder, and oppression. The continued support of understanding planetary relations as a choice between nature and culture only adds to the human tendency to separate ourselves from the so-called natural world.

State Centrism

I begin by unsettling these spaces of academic inquiry and those bodies the discipline has taken for granted with a discussion of the state as defined by the scholarly field and political institutions of international relations. For those unfamiliar with IR, this discipline is decidedly state-centric, with a focus on human institutions and structures. The state is the ur-concept that organizes thinking and method in IR. By this, I mean that the state, defined as a political community bounded by its territory and externally recognized by other states, with a population and government and with a monopoly on violence within its borders, figures centrally in IR discourse and practice. Whether it is ideational through constructivist theory or a priori for structural realists, the state is taken as the primary agent in

International Relations—as its fundamental ontological fact. Sovereignty, or the authority of the state to govern itself is expressed both through the state as a territorially bounded unit and through men and (sometimes) women as self-limiting, self-understanding, self-conscious, and self-representing subjects. The state's authority as the supreme power in the international system is both undivided and absolute. Internally, it can be democratic, republic, or monarchic, to name but a few ways sovereignty can be instantiated in various states. As codified in the Montevideo Convention on the Rights and Duties of States, a state must have a population, a defined territory, a government, and the ability to enter into agreement with other states. Internal sovereignty is always regarded in context with other states recognizing that sovereignty. Therefore, the international system is created through the mutual recognition of each of the units in the system.

In reflection of these commitments, much of mainstream International Relations writing has been concerned with the shifting and changing international system in relation to state power; a system that was perceived as relatively predictable and stable during the Cold War, but one that lost coherence after the fall of the Berlin Wall and the Soviet Union. The "end of history" was declared with a liberal victory over state communism.[20] Alas, this end point of history has not proved to be any less tumultuous and violent than previous eras.

In addition, changes in warfare techniques and acts of terror demonstrate that the state may be ill-equipped to respond to the asymmetrical forms of violence in the international system, and state reaction to international terrorism has likely created more problems and misery than it has solved. One has only to look at the legacy of the U.S.-led Iraq wars and the rise of new terror and insurgent groups in the region like ISIS. Climate change and refugee flows have bared weaknesses in the international order based on the territorial state to handle both complex humanitarian emergencies and issues of accountability for violence and its prevention.

Even with all these challenges, the state continues to hold our collective political imagination. To quote the *Economist* article from 1996, "The State is dead, long live the State."[21] Often, as Patrick Jackson identifies, much of the theorizing about the international state system centers on the state as an actor "without any systematic effort to theorise what it *means* for something like the state to be an actor or a person, and thus displays a lack of clarity on the question of how we would know whether (for example) the state were fading away or not."²² This debate over the presence and disappearance of the state has acted, as R. B. J. Walker posits, to depoliticize critiques that would direct us toward potential alternatives to the state—especially those that offer "something more human, global, or ethical, or environmentally sensitive."²³ As I frame it above, IR is often unable to see planetary responses even though it purports to be the most global of disciplines because its focus is on the ways in which the state is waning, in crisis, failing, or unable to handle the increased flows of capital, diseases, and people. As for the state, it does understand flows of capital and intellectual property as directed by corporate activity, but when faced with transnational issues like refugees, migrants, criminal activity, climate change issues, and pollution, it finds itself lost in a sea of complexity—a place of overlapping people and things that confound state power.

A main contention of this book is that the state-centric discourse of International Relations and lack of attention to other forms of political organization have led to a static conception of human lives within this system. Understanding relies on bordered thinking and a superimposition of Westphalian order on the human body. The discussion of human life, and human bodies, is largely through the language of state sovereignty and the Western, liberal subject, underwritten by ideas of the individual that are based on rational, economic actors or limited through realist assumptions about the fallen and wicked nature of the human condition. These discussions often address the individual through scholarship on human rights or through identity politics or ethnic studies, but the state-based focus of the discipline leaves the material body understudied and undertheorized. If IR theory and practice begin from the above assumptions and perspectives, the human body and life can be secured through national security practices, rational foreign policy decisions, global free trade agreements, or the spread of democracy as a system of governing. Any danger to the body merely needs more national security, or expanded trade regimes, or democratic institutions more closely aligned with the general will. These assumptions often do not examine how they may endanger the actual human body and limit the scope of responses that the international system may have to make to protect people and environments.

Take, for instance, economic development and its environmental consequences. Although electricity improves the lives of human communities, a hydroelectric dam project will displace thousands, and mining for uranium to power nuclear reactors will poison nearby communities with tailings. All told, capitalism is dangerous to our continued survival as a species, but it has also done much to decrease human suffering through wealth creation. As another example, take military intervention for humanitarian purposes. This broad case can demonstrate the dangers of getting trapped in the circular reasoning of a state-based option. Why is intervention needed? The state is breaking international law by withholding guaranteed human rights or committing violent acts against an ethnic group within its borders. The solution: intervene with violence and then reestablish the state. Elsewhere I have argued, with many others, that this is an irresponsible act that does not address underlying causes of violence found in the political foundations of the state.[24] The international community needs to reflect on the violence that comes bundled, so to speak, within the state as a political form and community. More on this later.

The Microbial State will explore the ways in which we create ourselves, both as individuals and as humans, beyond how the state predefines our identities as citizens. As discussed above, International Relations, by virtue of being shaped by its dominant theories, generally uses a tripartite level of analysis schema: this focuses attention on the individual, or the state, or the international system. Each is treated as a discrete subject with few or no ties to the other levels. It is state-centric and this state always fits within a system of states. This book hopes to disrupt this organizing logic of IR. In part, it will do so by reevaluating the actual and metaphoric composition of the human body and how human bodies might create a state, or a political community. It does this by using the metaphor of the body politic and processes inside the human body, its corporeality, as guides to create a counterstory to disrupt this state-based narrative.

Scientific Approaches

A commitment to multidisciplinarity brings the biological, body politic metaphor at work in the social sciences into engagement with developments in the life sciences to draw a more complete picture of the relations between diverse bodies and the possibilities that these understandings can open for the creation of a truly global politics. While research in the life sciences has implications for our understandings of the human body and interactions between bodies, it does not offer us easy answers based on biological reductionism.[25] Put another way, I am not proposing that there could be an essentialist underpinning of the "real" body underneath all of our "constructed" cultural interpretations. Instead, scientific discoveries continue to complicate and enrich our ideas of what it means to be human, and may offer a productive way to envision alternate foundations for social science. I am not searching for a biological explanation for social behavior, but rather want to find a more radical empiricism in each of the disciplines investigated.[26] Indeed, what is more real than the bodies that give us life?

The social sciences and the life sciences have much to offer each other. This book seeks to add to a burgeoning interest on both sides of the social and biological sciences to connect the disciplines. "There is a growing belief that our ability to address many critical social challenges can be transformed by greater knowledge of social behavior's biological foundations,"[27] and at present, few links exist between biologists and social scientists. William Connolly contends that there are diverse sources for this: the reductionist tendencies of biology and neuroscience and the difficulty with engaging these sciences from the outside and cultural theory's desire to create and protect its own foundations.[28] Often, it is academia and the modern university with its isolating fields of study into silos, each committed to its own epistemologies and competing for scarce internal and external funding. The biological sciences and the social sciences, often in the same colleges in U.S. universities, are separated into different departments with few meaningful connections between them.

This lack of dialogue is not a self-correcting problem, and an effort must be made to support "transformative biologically informed social research."[29] Multidisciplinary projects can create reflective spaces and benefit both scientists and social theorists by adding extra dimensions and debate to ethical and normative issues that arise with the advent of new technologies. Both philosophers and scientists are trained to analyze complex situations, and these different viewpoints can only add dimension to scientific and social knowledge following biomedical advances.[30] Moreover, modes of collaboration between the human sciences and the life sciences can open doors to new possibilities and understandings of the complex interactions between humans and nonhuman actors. This aids in breaking down dominant assumptions about the divide between nature and culture, and mixes in creativity and enhanced understandings of all forms of participation and agency in the world.[31] Not only can the social sciences benefit from a more rigorous knowledge of biology, there is also a need for scientists to stay aware of the social and political effects of their research. Craig Venter, part of the team that designed, synthesized, and assembled *Mycoplasma mycoides,* a synthetic cell, explicitly addressed the ethical and moral issues surrounding the creation of synthetic life.[32] "This is a philosophical advance as much as a technical advance," he said, and suggested that this raised issues about life and our understandings of it. At the time of this discovery, Venter employed social theorists as well as scientists to study all sides of this emerging technology—a synthetic genome whose parent was a computer.

My emphasis on the reciprocal nature of this relationship as a dialogue should make clear that I am not using science as a trump card that assigns to this book and its contents the role of unassailable truth based in fact: Science does not always have an unmitigated view of reality, truth, and the "world as it is." Scientific practices are not removed from the social and political and there is no pure science that is not influenced by the social world. Examples from the life sciences are not used to "fix" the mistaken ideas of social science, nor do I argue that the life sciences will suddenly make clear what is now muddied in political and social interaction. It is not a focus on the changes that science fuels, but on what these changes point toward and how we can use them to understand the way in which power circulates in these statements and institutions.[33] As Bruno Latour stresses, there is no "magical machine that speaks the truth and pays no price for it in controversy."[34] Neither IR nor science are hegemonic; they include sets of practices that differ over time and space, the desires and directions of individuals, and varying and diverse epistemological commitments. A dialogue between the two brings out the complex nature of who decides, defines, and controls who is a subject, what is an object, and how each are treated politically and ethically. These discussions should be open, inclusive, and careful to reflect the values we feel are necessary in creating mutual public space.

The two examples I draw on from the life sciences to support this conversation and the shift to a more "flexible and appropriate stock of metaphors that can generate fresher and more enlivening analogical accounts"[35] are metagenomics—genetic analysis applied to entire communities of microbes and studied in a way analogous to a single genome—and new research into the nature and functioning of the human body's immune system. Specifically, the field of metagenomics has revealed complex communities of bacteria that inhabit the intestinal tract of humans. These bacteria protect the human body from disease as well as digest and metabolize food that humans could not otherwise process. In fact, the DNA of other life forms in our body outnumbers human DNA ten to one. I use this to generate ideas of community that allow for a politics that is committed to supporting relations within states and across borders as the human body does between many microbial inhabitants.

Increased understanding of the role of the immune system is explored in the second example. I argue that new dimensions in immunological theory, as well as our lack of humanity at the genetic level, may advance some of the most contested issues in the history of Western thought such as the nature of the Cartesian self, the role of identity creation and difference, and how we differentiate "self" from "other," or nonself. The human body, through its immune response, can give us clues to support life in such a way that does not discount or endanger otherness and difference in ways that simultaneously go beyond the human and involve a rethinking of what it may mean to be human. Using these examples, my concern is not to "deepen" or discover a more "authentic" meaning in the body itself, but to investigate how body metaphors work to justify or excuse acts of state violence and how they can aid in addressing this violence. I show, for example, that simplistic metaphors likening illegal immigrants to viruses attacking the immune system invite violent responses that are justified through a discourse of "health" for the body-polity. An expanded and more complex understanding of internal state health disrupts this easy justification.

In other words, the traditional figure of the human body—as a self-contained and self-regulating organism—is at odds with the body made possible by new technologies and understandings. Today, organs and genomic information flow across borders, and bacterial and viral communities, both symbiotic and pathogenic, clearly affect our bodies, and through our bodies, politics. It is becoming more and more difficult to uphold the fiction of autonomous selfhood, or the figure of the human body as a Newtonian entity with boundaries. A larger question that the book engages with is what it means to be human in a world complicated with a multiplicity of life forms and an increasing reliance on technology. As Timothy Morton comments, "It might be important to figure out whether persons really are solid, single, lasting, and independent beings. This has huge implications for ethics and politics."[36] Through these examples from the life sciences and how they complicate our social understandings, I question the nature of what it means to be human, and interrogate how we interact at the corporeal level with other humans and nonhumans. Therefore, a specific and central aim of this book is to apply sustained critical pressure to the autonomous and autarkic ontology of the individual and the state as currently defined in international relations.

To aid in this critique, the book creates an analytical structure able to identify and celebrate plurality without erasing internal diversity, or encoding the external as strange and dangerous to an imagined unity within. Born from these examples is a pair of novel metaphorical framings to help build a different conception of humanity's myriad ties to the world: the *lively vessel* and the *contaminated state*. These place bodies in relation to other bodies, rather than individual survival or self-interest. Contrary to many theories of security, increased safety and welfare can come from being more connected, not less. While modernity has meant, in part, a focus on the individual and the individual's protection vis-à-vis the collective, lively vessels and contaminated states provide new metaphors named for the processes that intertwine multiple bodies into composite ones different from the current definition of individuals, nations, and states as self-reliant and self-defined.

Theoretical Approaches

This book is, by necessity, a conceptual and theoretical work, but it is grounded in the empirical: "good concepts serve to validate and highlight empirical detail, and reliable facts are needed to corroborate and expand explanatory concepts."[37] It follows Paul Feyerabend in his assertion that the prioritization of theory over experience and vice versa is often specious: "Experience arises *together* with theoretical assumptions not before them, and an experience without theory is just as incomprehensible as is (allegedly) a theory without experience."[38] It is also the case that there can be no simple return to empiricism or positivism as the debates have clearly moved beyond the subject/object distinction. It is a world ruled by post-Einstein physics and complex physical and natural systems—a contemporary context that demands engagement with the "real."

To bring this complexity to the fore, I draw inspiration from a school of thought often referred to as *critical* or *new materialism*. It is labeled new not because of its place on a temporal or teleological line, but due to the "unprecedented things that are currently being done with and to matter, nature, life production, and reproduction." This newness compels theorists "to rediscover older materialist traditions while pushing them in novel directions or toward fresh applications."[39] Materialism gives matter and nonhumans, like bacteria, agentic capabilities by bringing the object into a place of its own not just as related to the knowing subject—or human; this focus refuses to downplay the nonhuman as something static and only brought into being by its encounter with human consciousness.

New language and metaphors can begin to form and emerge if they are added to the idea of IR as a site of complex cultural practice.[40] IR can then become a forum where beliefs about the world—its composition and organization—are brought together, debated, produced, and reproduced, and therefore it can be an ideal place to pursue the questions outlined above. IR theory has assembled various ways to bring meaning to the world, and the practitioners of each theory, through their implicit and explicit ideologies, have put forward a coherent and comprehensive set of ideas to explain the world. The various theories at work in explaining International Relations and global politics aid in understanding an individual's place in the larger picture, and create a program for political and social action.

This program for action is aided by the close assessment of the metaphors born from the experience of the corporeal body, and the politics these metaphors hope to naturalize. This means interrogating what metaphors *are for*, and not necessarily what metaphors *are*.[41] Put differently, this is not an exercise in discourse or linguistic analysis, nor is it an attempt to contribute to an already large and existing literature on metaphor in politics and political language. Instead, it hopes to define how bodily metaphors can be productive,[42] both as a way to elucidate and acknowledge the ways in which bodies matter—in spite of the undertheorization and gaps in the literature—and to identify how understandings of the actual body will add to debates about the actors acknowledged in international relations. Metaphor plays the role of interlocutor and offers a way to organize my arguments about the creation of politics centered on the body.

More generally, this engagement opens a dialogue between the social and biological sciences that serves two purposes: first, it hopes to illuminate the work that metaphor does within the social and the life sciences by showing the consequences of using certain metaphors over another; and second, it highlights the creative power of metaphor to change our understandings of the world. This is not to argue against the power of metaphor as such, or to insist that we should move away from metaphorical and figurative language. In fact, I argue the opposite. Metaphor is a necessary and unavoidable tool for understanding the world; it takes abstract concepts and makes them comprehensible through bodily experience, cultural definitions, and cognitive processes. This means that much of what we do everyday engages in metaphor and metaphorical thinking. Importantly, metaphor is a conceptual and contextual system as well as a force with physical effects on the human body; it often plays a much more significant role than fanciful wordplay or poetic language.

To connect this power of metaphor to bodies, the book draws upon, and extends, George Lakoff and Mark Johnson's work on cognitive metaphor where "understanding emerges from interaction, from constant negotiation with the environment and other people."[43] This works through "mapping" concepts from a source domain to a target domain. Source domains come from the physical world and are often things that are easily understood or experienced bodily. Target domains are the conceptual and abstract worlds of ideas, mental or emotional states, and social understandings. Simply put, we take our experiences and understandings of the physical world and use them to explain abstract or nonphysical phenomena as something we can relate to from our material experience. Used in the natural sciences for similar purposes, metaphors help explain the "unseen and unknown domains of the physical world as, for example, the world of molecular action."[44] They are "conceived of as crucial tools that contribute to conceptual and theoretical innovation in the natural sciences" that "facilitate the production of concepts or theories that would not be available otherwise."[45] Then, in turn, metaphors induce changes in the physical world. The strength of metaphor comes from and returns to the material world of bodies.

The primary metaphors highlighted are ontological and conceptual. These take unseen concepts that can be defined by experiences with physical objects, especially our own bodies. Lakoff and Johnson define ontological metaphors as "ways of viewing events, activities, emotions, ideas, etc., as entities or substances."[46] These ontological metaphors "allow us to make sense of phenomena in the world in human terms—terms that we can understand based on our own motivations, goals, actions, and characteristics" (34). Love can serve as an excellent example. It cannot be measured as such, but we describe it as a bond, a flame, a journey, or a garden. Love is in the air and love is a battlefield.

These understandings of unseen things often rely on spatial orientation metaphors such as up–down, front–back, center–periphery, or near–far.

Nonphysical ideas can be "seen" or felt as objects and understood in orientational terms stemming from the fact that we have bodies that function in certain ways in our environment (14). For example, being happy or sad is referred to in terms of being up or down. "I'm feeling *up*. That *boosted* my spirits. My spirits *rose*. You're in *high* spirits . . . I'm feeling *down*. I'm *depressed*. . . . My spirits *sank*" (15). It is plausible that this came from a certain physical basis: "Drooping posture typically goes along with sadness and depression, erect posture with a positive state" (15).

Along with the bodily and ontological nature of metaphor, which keeps the examples from being overly focused on what metaphors are defined as and not what they do, I insist that metaphors and metaphorical framing can be *innovative* and *generative,* as well as descriptive. "I have said before that metaphors are dangerous," writes Milan Kundera in the *Unbearable Lightness of Being*, "Metaphors are not be trifled with. A single metaphor can give birth to love."[47] If metaphors have consequences beyond being "just" rhetoric, then it follows that questions should be posed about how metaphors may "form/deform the experience of those making policies."[48] They have *both* linguistic and nonlinguistic efficacy: they can justify acts, authorize certain actions, inform political decisions, and create a milieu, or a framework, in which actions are taken. This creates potential for both abuse and transformation. In his analysis of the first Iraq War and metaphor, Lakoff writes, "metaphors can kill."[49] The metaphorical frames supporting action in Iraq, for example, that of the evildoer, the victim, and hero, led to the justification of particular actions over others. Susan Sontag, in her book *Illness as Metaphor,* writes "Metaphorical trappings deform the experience of having cancer" and delay treatment. "The metaphors and myths, I was convinced, kill."[50]

The focus on metaphor will function in two ways. First, I use the body politic as a tactic for dialogue in a political-theoretical discourse that, in an ironic and productive disavowal, does not take the body as a legitimate object of study but does use the metaphor readily in its literature. The body politic is an overarching signifier that can ground and contain discursive practices and interplays between theory and praxis, bodies and politics. Second, I draw on the figurative strategy and linguistic meaning of metaphors more broadly—specifically, the power of synecdoche—to make a concept, an object, or a thing stand in for another. This can be more powerful than a simile, in that meaning is radically changed and intensified: it is not "like" the other thing—it becomes the other thing. In the second chapter, I stress the *metaphoro-genesis* of concepts and the effectiveness of metaphorical framing, or *metaphoricity*, in influencing our opinions and judgments about the world. Metaphoricity is defined as the power of a metaphor or, put another way, as our awareness of its being a metaphor. A dead metaphor is one that has turned into a concept with no memory of its once having been figurative. Nietzsche provides excellent guidance in this regard:

> What then is truth? A mobile army of metaphors, metonyms, and anthropomorphism—in short, a sum of human relations, which have been enhanced, transposed, and embellished poetically and rhetorically, and which after long use seem firm, canonical, and obligatory to a people: truths are illusions about which one has forgotten that is what they are; metaphors which are worn out and without sensuous power; coins which have lost their pictures and now matter only as metal, no longer as coins.[51]

Metaphor, when exhumed, has tremendous creative and fixative power, and thus political and social power. There is no need to worry about truth and its illusions, but rather to remember, to feel the possibility in the sensuous power of the body politic as metaphor. The body should be more than a dead metaphor.

Indeed, theory and metaphor aid in bringing out feelings about the world to contribute to a larger language of affect about whom and what we share worlds with on this planet—affect not just as attention to our body and emotions, but to what (and whom) acts upon us. This is a view of causality that assigns affect to both sides of a causal relationship, or, in other words, a Spinozan sensitivity to the pull of other bodies; a processual concern over what bodies do rather than the Kantian concern about what we can know about these bodies; a Derridean experiment about singular beings and their ability to "live together" in a "democracy to come."[52]

Poetic Counsel

To bring this affect grounded in material entanglement and poetic critique of the status quo to the fore, the chapters begin with lines chosen from *Leaves of Grass*, Walt Whitman's book of poetry. They provide a creative subtext to the analysis done in the chapters, a way to open an interest in welcoming the stranger and to invoke a material openness to the corporeality of the world—a way to feel the sensuousness of the body politic. Whitman wrote that poets were best suited to "strengthen and enrich mankind with free flights in all directions not tolerated by ordinary society." Poets know no laws but the laws of themselves, and are not beholden to "mere etiquette." Whitman said, "Often the best service that can be done to the race, is to lift the veil, at least for a time, from these rules and fossil-etiquettes."[53]

Walt Whitman's project was one of cultural and literary revision against the prevailing notions of the body and its relation to politics and society. He was intent on "rewriting notions of what the consequences of embodiment or the meanings of a body, its parts, functions, products, and states may be."[54] Whitman produced texts that extended his reader's conceptions of the body and the literary, and especially how these categories interact to exceed or overrule the cultural constraints of the time. Through *Leaves of Grass* and its many revisions, Whitman joyfully supported the body as a fluid self who struggles to negotiate identity and difference while committed to being responsive to as much of the world as possible. "He pushes us to make room for variety in all its varieties, a project whose radical force is reigned in and domesticated in most calls for inclusivity, which usually meant inclusivity of a finite and preferential range of social phenomena."[55]

With Whitman as poetic counsel, I approach the topics in *The Microbial State* with love and care, but also with a conviction that seeing possible alternative global orders is of the utmost importance. I hope to refresh a belief in the importance of plurality as a critical condition necessary to being human. As Hannah Arendt writes, "Plurality is the condition of human action because we are all the same, that is, human, in such a way that nobody is ever the same as anyone else who ever lived, lives, or will live."[56] This may enable us to build a richer and more inclusive respect for life in global politics, knowing full well that there is no one option that makes right that which is wrong with the world, but nonetheless we must respond.

Whitman offers a model for "cosmopolitan social imagination, deep pluralism, wide-ranging habits of contact, and attention to corporeal individuals" that can "serve as useful equipment for new rhetorical and moral production."[57] This response may not be as an actor who identifies a problem and then "fixes" that problem, but it creates awareness that humans are part of the problem itself, and as individuals we are likely to be party to many of the crises we are responding to globally and locally. Therefore, an ethos of care for the world is crucial, and ethical engagements need to keep these points in mind.

Positive Biopolitics

I conclude with some final points as to why this method and its ethos offer a promising way to interrogate the structures and institutions of global politics. It remains important to offer creative and disciplined thinking about the relation of life to politics in the international. Too often the discussion centers on negative instantiations of biopower, or technologies that were created for managing populations with life at the center of concern, and on the biopolitical control that is expressed in global institutions and relations. This is often a negative emphasis, focused on the body as in "a liminal state of the extreme vulnerability of being human: becoming-corpse."[58] I argue that biopower is to "be grasped not merely as the capture of life as an object of power" but as "intensive, creative, and infinite in its Spinozan take, in which life became a subject as well."[59] This reversal of biopolitical critique, one that emphasizes vitality connection and entangled responsibility, is an underlying theme of all the chapters. This book takes life as a creative intensity that can offer new solutions, and new ways of engaging with the world. This will be the task taken up in chapter 4.

Placing an idea of life as vital at the center of politics could lead to two important implications: first, a rethinking of ethics and responsibility leading to a "diffuse sort of ontological gratitude for the post-human era, towards the multitude of nonhuman agents" that support us.[60] This diffusing, or flattening, of social action and ties into a continuum of dynamic object interactions, or translations, between humans and nonhumans, states, bacteria, biomes, and parasites, makes the nested and imbricated nature of politics inside and between body politics more visible.

The second implication is explicitly political: Human politics needs organized collectivities, institutions, and organizations to reflect these "dreams" of nested subjectivities. As Latour asks, "Once the task of exploring the multiplicity of agencies is completed, another question can be raised: What are the assemblies of those assemblages?"[61] To go further, I ask what could, or should, the assemblages of our assemblages be, with a humility that appreciates that in the Anthropocene "we" humans and states cannot answer this question as freely as we thought. These discussions should be open, inclusive, and careful to reflect the values and ethics we feel are necessary in creating mutual public space and a "critical proximity, not critical distance, is what we should aim for."[62] This is a matter not just of inclusivity, but survival.

Human bodies, and the ethics and politics that follow from this engagement, are brought into conversation with the nonhuman. This will open the reader to the presence of "things," and nonhumans, at play in our bodies and politics. This "becoming sensitive," as Latour calls it, can teach us lessons both about international relations and human relations, as well as give us a wider understanding of planetary existence in which humans and nonhumans are co-constituted. How are we called to respond or to be sensitive to beings in the world—be they human or nonhuman? The conversation begun in this chapter cannot happen without the life sciences. It is therefore imperative that we support the idea that science and politics are not adversaries, or removed from one another, but are participants in a collective process of world-making. Rather than humans being bound to the earth as a prison, the life sciences can aid us in connecting to the earth as a place that supports life and gives us relation to all other species, for, as Arendt writes, "the earth is the quintessence of the human condition."[63]

This collective world-making must nurture positive attachment to the world and consolidate a belief in the world beyond its use value. This is not just a belief that supports the established regimes of power and their institutional instantiation. In fact we need "to resist and overcome with positive alternatives," and find and accumulate positive attachments to the world. These attachments must understand that we live in a world where *"numerous constituents we encounter on a regular basis increasingly bring different conceptions of the world to experience as such."*[64] This focus on pluralism encompasses more than a commitment to a secular or liberal multiculturalism as it entails more than just an acceptance of diversity and a tolerance toward that which is different. Supporters of pluralism must engage with energetic interest across lines of difference and, importantly, support differences as if their own lives and freedom depended on it.[65] How do we find a way in our politics to create a deep pluralism that can take these constituents and conceptions into account?[66]

One such way might be to build new understandings of global practices that recognize the "vital and shared" materiality of our human constitution and furthermore "to conceive of these materials as lively and self-organizing, rather than as passive or mechanical means under the direction of something nonmaterial, that is, an active soul or mind."[67] As humans, we must live, as Donna Haraway, feminist, theorist, and scientist said, in the "concatenated worlds" that we have inherited through our connections and encounters in the world.[68] To admit, along with Lynn Margulis and Dorion Sagan, the biologist and her son who theorize and write on how our growth as a species happens as a result of coevolution with multiple other species, that "the completely self-contained 'individual' is a myth that needs to be replaced with a more flexible definition."[69]

One can see that many contested issues in the history of Western thought, such as the nature of the Cartesian self, the role of identity creation and difference, and how we differentiate "self" from "other," or nonself, can be answered differently. Are there better metaphors that can be used to understand the body and its relation to its "outside" and "inside"? This will mean extending the subject and objects of human politics in new directions and accepting multiple layers of agency. These extended metaphors created by corporeal thinking and their lively effects and affects are the subjects of what follows.

This will include the ways in which states, society, and organizations *compose* themselves as larger bodies analogous to the biological subject and how the *constitution* of these real and metaphorical subjects is narrated, performed, and made real in global politics. This focus on constitution will demonstrate how many of the dominant paradigms in IR are inadequate to respond to the current biopolitical, industrial, and advanced capitalist regimes of power currently at work globally. I argue that it is imperative to think "about matter, materiality, and politics in ways that do justice to the contemporary context of biopolitics and the global political economy."[70] The kind of body politic, or bodies politic, that can be created from these musings are the topics of the next chapters.

1

Corporeal Politics

*Does the earth gravitate? Does not all matter,
aching, attract all matter?
So the Body of me, to all I meet, or know.*

—Walt Whitman, *Leaves of Grass*

We are all bodies, but we are often not aware of how corporeality matters. Our bodies are frequently understood not as material, but rather as ciphers or discursive constructions to be filled with meaning, both biological and social. We are defined politically by our citizenship, or our legal personhood, or as a voter, rational actor, or consumer, but not often as an actual body that can be fragile, leaky, diseased, sold, colonized, male, female, multiple. This body is a powerful body with mysterious impulses; a body that cannot be fully regulated, controlled, or precisely monitored. In this way the body can be a material reminder that politics, culture, and biology cannot completely describe our corporeal forms or our experiences as a body among others. Bodies—and life—exceed and resist our definitions. This bodily remainder is often captured as metaphor: we use the power of the body's corporeality to define and inform politics through language.

A desire to know more about the material body and its relation to our social, political, and cultural systems has led to increased curiosity about the body in varied disciplines, but International Relations (IR) and its focus on the state has conducted limited scholarly investigations into the body.[1] Nonetheless, theories, ideas, and discussions about the individual body, its claim to life, and its place in the discipline can be found in various recent literature. One can find bodies in the very words of international relations: organs of the United Nations, the family of nations, and heads of state. The people and their nation, too, have been envisaged as unitary bodies subject to uniform systemic or constitutional processes or functions such as democracy and the social contract. There are dead bodies, elided in the debates around military and civilian deaths during war ("we don't do body counts"[2]) and cared for in the laws surrounding the bodies of fallen soldiers and (some) noncombatants in the Geneva Conventions; bodies in motion as migrants and refugees; dangerous bodies in the form of terrorists and *genocidaires;* and bodies in danger because of state failure, natural disasters, and crime. Even with the body as a powerful shadow falling across a range of IR topics and concerns, the body as a situated and material entity in its own right is abandoned at the edge of the discussion—depoliticized and powerless. The rationalist, individualist paradigms in IR tend to ignore the "importance of situating empirical actors within a material environment of nature, other bodies, and the socioeconomic structures that dictate how they find sustenance, satisfy their desires, or obtain the resources necessary for participating in political life."[3]

Interest in the body, and its seductive power to hold IR's attention, is unmistakable through the long-standing debates about the state's personhood. Obvious care for the body is evinced in human rights law and the laws of war. It is reflected in the debates over intervention for humanitarian purposes, individual accountability for war crimes and crimes against humanity that have supported the creation of the International Criminal Court. It is further evidenced by metaphors like the "body politic" and its continued presence in the canon of IR as an organizing principle and ontologizing fiction. Following in the footsteps of Machiavelli, Hobbes, Kant, and other precursors, for instance, IR theorists have understood the state as a person—an "artificial man" and a "Body Politique."[4] It is undeniable that bodies matter in IR, but in what way? This book grapples with that question.

In any case, the question arises again as to how the actual body is understood and represented in IR and to what ends? How does it matter? Although this book cannot offer—and, indeed, would not want to offer—a definitive answer to this question, it explores the possibility that a focus on the corporeal body can suggest productive responses to the puzzles of global politics today. Michel Foucault, to the editors of *Quels Corps?* said, "One needs to study what kind of body the current society needs."[5] This chapter takes as its focus a similar question: What might a corporeal politics look like globally?

To this end, what follows is an exploration into the visceral and metaphoric power of bodies to address global politics. This is not the individual, the citizen, or the person as IR understands it, but a focus that rests on a "corpus." A corpus, Jean-Luc Nancy writes, as a collection of pieces, bits, members, zones, functions. Heads, hands, and cartilage, burnings, smoothness, spurts, sleep, digestion, goose-bumps, excitation, breathing, digesting, reproducing, mending, saliva, synovia, twists, cramps, and beauty spots.[6]

With this focus, I need to address IR's disciplinary understanding of the "proper" objects and subjects of study. How can the body inform a discipline that takes as its main object—and subject—the sovereign state? Indeed, it can be argued that IR has chosen its objects/subjects of study from a rather short list. By using both subject and object, I am intentionally playing with the larger philosophical debate about the Cartesian subject and the separation between an immanent subject and a transcendent object, or the extension of mind versus the extension of matter. This will be taken up in more detail in subsequent chapters.

Politicizing Bodies

IR discusses the State with a capital S: it is a discipline defined by its theories of the state. In the years between World Wars I and II, after World War II, and again after the fall of the Soviet Union, IR writes its history as one that is organized by the "Great Debates." These debates tell the story of a discipline reacting to and changing according to the material conditions of the world around it.[7] Admittedly, more and more attention is focused on the individual, but it is generally bounded by international law or rationalist assumptions of individuality over sociality. The International Criminal Court and international human rights regimes have brought individual accountability to a system of sovereign immunity, but progress is slow and fitful. In addition, IR is decidedly wedded to its levels of analysis: the international system, the sovereign state, and the sovereign individual. The hierarchical internal state and the external anarchical nature of the international are the organizing logics for understanding conflict and interaction in global politics and international relations. To add to this, the state understands the international through an immunitarian logic that polices and protects the inside from the outside and its dangers.

This book takes a different approach in that it shifts away from a singular focus on the state, or more precisely, it does not take the composition of the state as given, but rather desires to explore what is *un-stated* about the state.[8] What underlying processes are taken for granted? How do IR's assumptions about the nature of the state and the anarchical international system that disciplines it obfuscate other processes at play? What goes without saying when IR speaks of the state? In other words, the ontology of the state (among other states) is not understood as existing a priori, but rather analyzed a posteriori as complex and interrelational, virtual and material, as both object and subject.

These interrelations are buzzing with things, with objects, with bodies of all kinds and shapes. Bodies that have multitudinous needs and desires. In this approach, I create an interdisciplinary "cabinet of curiosities" in which these things can be reformed into matters of concern. This is not just an extension of "subjecthood" to other international actors that have simply not been found or recognized yet, but rather a genuine curiosity about how a future politics may be organized differently; it is about how things, or objects, "need to be represented, authorized, legitimated, and brought to bear inside the relevant assembly."[9]

At the same time, it must be remembered that the state is not a thing, or corpus, in the above sense of the word, nor is it an object composed of matter. Unlike the human body, which can be studied as an object, it cannot be seen as a whole or studied empirically like other objects. The state is a great imaginary—a powerful conceit—created to organize a community politically and socially. The state has to be made real; it is invisible until it is performed, or as this book will remind IR, until it is made visible through metaphor. A metaphor gives reality to what cannot be seen through the ontological and conceptual forms of metonymy, synecdoche, and allegory. It is an answer to the question of political order rather than the question itself. It is for this reason that the book uses the dual structure of materialism and an object-oriented focus coupled with the ontological metaphor of the body politic. This is not to say that it does not wield power and capacity to damage; the state's most important performance is that of the Weberian "monopoly of violence" over all that exists within its borders.

I am explicitly playing with the limits of the dichotomies already present in the debates: language and affect, the virtual and material, and the object and subject. In *Making Things Public,* Latour points out that Hobbes was not unaware of the material world in which this Leviathan resided:

> the "Body Politik" is not only made of people! They are thick with things: clothes, a huge sword, immense castles, large cultivated

fields, crowns, ships, cities and an immensely complex technology of gathering, meeting, cohabiting, enlarging, reducing, and focusing. In addition to the throng of little people summed up in the crowned head of the Leviathan, there are objects everywhere.[10]

After considering these broad themes and reflecting on both implicit and explicit bodies in IR, a question can form based on a conception of a body that is also important to the discourse of politics: what about the metaphorical body, or the *body politic*? The human body enjoys a long tradition as a metaphorical framework for political and societal organization in classical (and contemporary) political theory. The birth of Hobbes's *Leviathan* helped to create and develop modern forms of government, industry, culture, and the nationalisms that we know today. The metaphor of the body politic is the fusion between sovereign and subject, but how does IR understand the interplay between metaphor and actuality and can it aid in exploring the question of a future-oriented corporeal politics?

Body Politic

To begin, the reader should consider the detail in the frontispiece of Hobbes's *Leviathan* (Figure 1). Before one reads a word, this picture tells a story about bodies and politics: a man holding a sword and staff looms over a city; these represent the sovereign's control over the world of earthly politics and his right to decide upon the next world of the soul. He wears a crown, and his expression is both benevolent and withdrawn. His shadow falls upon the countryside with farms nestled in the hills. Upon closer reflection, we see that the man's body is composed of smaller bodies, men within a man.

This is the sovereign: a visual representation of the "body politic," or a political community imagined in corporeal form. Hobbes ties the natural and artificial body into one through this imagery. In the Leviathan, he explains, sovereignty is the soul, and magistrates are the body's joints, reward and punishment, its nerves. Money is the blood providing nutrition from the Plenty of Nature; and its colonies and plantations are his children.[11]

Further, the body politic is a term that hitches our corporeal experience to the practice of politics. In Western tradition, the idea of a collection of bodies forming larger bodies, like *Leviathan*, is common and is most often

Figure 1. The frontispiece of the book Leviathan *by Thomas Hobbes; engraving by Abraham Bosse.*

expressed as either the idea that members of a body politic are a voluntary community, or that a body politic is natural and emerges when a people perform their proper function, just as a cell would in a larger organism. From classical to modern thought, from Plato and the polis as "the body writ large" to John of Salisbury and the "organic analogy," the body as metaphor is a long-standing trope in political theory.[12]

As pictured above, *Leviathan* was an artificial body created in the image of man and made for the protection and governance of men. The body metaphor is present in Rousseau's *The Social Contract,* through the lifeblood of sovereignty, just as it is in Locke's *Second Treatise,* when Locke says that the body politic is created in the passage from the state of nature, which consists of many bodies with no common law or authority, to one body. Moreover, the body as metaphor is, in part, how subjectivity binds sovereignty and state. Anthony Burke explains:

> Hobbes and Locke laid out the discursive limits and conditions for the citizen as a form of subjectivity and bound the citizen to the state as an essential figure. Sovereignty became not merely a juridical foundation for the state as a concept and set of institutions, but a rhetorical trope that persuades the "citizen" of the state's inevitability, necessity and superiority.... All this reposed on a powerful political humanism, centred on the body, in which state and citizen find their identity and ontological ground—a circular movement, which begins with the liberty and reason of men in their singularity, imagines the state as the common body-politic of men in their collectivity, and returns as an enhanced promise of individual freedom within the now safer bounds of the state's supreme rationality and protective violence.[13]

This promise of security and belonging continues to provide sustenance for theories of the state; the body as metaphor is a powerful rhetorical image that circulates in politics and policy, affecting how people and states assume themselves to be secured. This story of state creation as an inevitable journey from the dangerous state of nature to the safety of civil society, or this fusion of many bodies into one body, whether the nation or the state, can legitimate violence and conflict as unavoidable and natural.

A main concern of *The Microbial State* is to investigate this journey that metaphorically fuses sovereignty and individuality. This focus then creates a host of cascading questions to explore: Should the body, as society and science understand it today, affect the metaphor, and conversely, how does the metaphor change politics that affect bodies? If human bodies are not the subjects they were once imagined, does this necessarily question current forms of collectivity created to protect the freedoms of these subjects? Conversely, can a critical accounting of metaphorical bodies and people aid in protecting actual bodies within the state from threats and insecurities? I believe so. The body politic has become fundamental to political life or, at the very least, is a lasting and powerful way to imagine the organization of political community; therefore, it is necessary to understand how the body is constructed socially and politically through its physicality. In other words, the body in the metaphor must be accounted for in theories of the state.

Here we can return productively to Hobbes. Four hundred and some odd years later, Hobbes's artificial man would likely have different forms filling its body. What would populate the body of this great Leviathan now? Because of continuing developments in science and technology, the human body and its mysteries are different than imagined by earlier Western philosophers, social theorists, and physicians. For example, X-rays, microbiology, and medical procedures such as organ transplants, make for understandings of the human body that differ from those of previous authors who wrote of the body politic.

This is not to say that there was not emerging scientific knowledge during Hobbes's lifetime. It is well-known that Hobbes and Robert Boyle disagreed over the political consequences of knowledge creation based on controlled laboratory experiments and the credibility of facts proven by experiments.[14] Hobbes and Robert Hooke corresponded, and it is likely that Hobbes read Hooke's *Micrographia,* published in 1665.[15] In addition, Hooke worked with Boyle in the laboratory. In 1676—four years before Hobbes's death—Van Leeuwenhoek saw "tiny organisms in water," so it is also likely that "minute bodies" and their physiology were a part of Hobbes's thoughts later in his life. Furthermore, three democratic revolutions—French, American, and Haitian—shifted sovereignty from God and monarch to the citizen and the territorial state form, making the Peace of Westphalia a watershed moment in European political organization. This European idea of social order and political space was transplanted from its origin to the rest of the globe through processes of colonialization and exploitation over the next three hundred or so years.

The remainder of the chapter builds the necessary vocabulary and introduces theoretical and metaphorical approaches to the questions of the body as I have laid them out. As with the state, the body and body politic are not always already there or solidified as a way to represent a mysterious presence otherwise not seen: they are created and performed. Speaking to the world, and its bodies, in a way that can attend to the material and agentic capacities of bodies will bring dynamism to discussions of global politics. This can be accomplished through attentiveness to the experiences of bodies and the plural formations that they can take.

Pluralizing Perspectives

This attentiveness is made easier by thinking of International Relations as a collection of stories about how we understand and engage with the world. It is also, both as theory and praxis, a collection of responses about how we create community and space for politics globally, and how we, as humans, survive in a complex, diverse, and finite world. Therefore, International Relations is, perhaps most importantly, about relationships—the creation, maintenance, and defense of the connections formed in this world. In mainstream IR, relationships are understood through the logic of anarchy and the security concerns of the nation-state.

These logics are underwritten by Western concepts such as "rationality" and "objectivity" that are universalized through the control of forms of knowledge within IR discourse. These discourses vie for what is "real" and "true" in global politics, and often pose as neutral explanations about how the world "really is." This is especially true in Anglicized and American IR theory, where a certain kind of political realism and positivism has debated with other paradigms over what is allowed to be true in matters of international life. Approaches from outside the West—geographically and theoretically—and IR's great debates pressure the mainstream approaches to open their theoretical and actual borders to ideas and approaches that better reflect global demographics and concerns. Hence, IR theory as a discourse is broad, contradictory, and often argumentative.

My interest in bringing certain scientific discourses into IR reflects an intellectual and disciplinary concern with further opening the field of IR to a plurality of approaches. This, in turn, supports a political and emancipatory program of "detaching the power of truth from the forms of hegemony, social, economic, and cultural, within which it operates at the present time."[16] As will become clear in chapters 2 and 3, this book can work in concert with "less West-centric foundations" that are "more respectful of multiple ways of understanding our complex world."[17] This interest supports the desire of critical approaches to be sensitive to self-reflection, reflexivity, and intersectionality within International Relations. Following Robert Cox's statement that "theory is always for someone and for some purpose,"[18] and also by noting that the questions and problems that drive theory are often more important than the solutions they propose, the book's framework critically approaches IR with questions and problems in mind rather than answers or an interest in "Truth." In this, it is helpful to remember truth's relation to power and epistemology. Foucault is essential in this regard:

> Truth is a thing of this world: it is produced only by virtue of multiple forms of constraint. And it induces regular effects of power. Each society has its regime of truth, its "general politics" of truth; that is, the types of discourse which it accepts and makes function as true; the mechanisms and instances which enable one to distinguish true and false statements, the means by which each is sanctioned; the techniques and procedures accorded value in the acquisition of truth; the status of those who are charged with speaking what is true.[19]

This leads to a critical stance that seeks to question these problems as well as our assumptions about the world and how IR frames it, and our response to these problems and assumptions. Gilles Deleuze writes that one must look at the problem to which the theory is responding; a focus on solutions does not serve when thinking philosophically. "People say to them: things are not like that. But, in fact, it is not a matter of knowing whether things are like that or not; it is a matter of knowing whether the question which presents things in such a light is good or not, rigorous or not."[20] Also important are the conditions of possibility for these questions; one must begin to see how things might be different if new questions are asked. There are no critiques of solutions, only of problems. Often, then, it can be seen that we have answered questions in a certain way and subsequently forgotten that a question was uttered in the first place.

Political Orders

The Westphalian state system is a good example of truth regimes and answers to previous questions of political order. The state, as we know it, was a way (an answer) to curb a particular form (question) of political violence in Europe between secular and sectarian forces. The Peace of Westphalia in 1648 is often recognized as the beginning of this sovereign state system as we know it, although it is another two hundred years before this system is more than a sum of its parts.[21] This state and state system—once a specific answer to a specific series of problems—has become universalized, the only way to imagine political community in the modern world. The questions of political order that states and the systems of states need to respond to have changed since the seventeenth century and they continue to do so.

Admittedly at the global level, this is a difficult task because there is no "outside" in which to predict, study, and reflect on what happens "inside." This lack of distance can cause feelings of panic as well as methodological conundrums. Importantly, as Timothy Morton stresses, "We are now compelled to achieve ways of sorting things out without the safety net of distance, ways that are linked to ways of sorting things out ethically and politically."[22] We are embedded in our world, and our thinking in the world and acting in the world are impossible to disentangle; therefore, attempting to untangle the relations that hold the world together would be a step in the wrong direction in our ability to attend to our ethical obligations. This would obfuscate the politics that will be needed to engage in the world. Or as Latour writes: "But obviously, in insisting ceaselessly on the existence of an external world beyond discussion, directly known without mediation, without controversy, without history, they render all political will impotent. Public life is reduced to a rump of itself."[23]

This is not to say that there is no outside world, no real world. This merely means that the approach and examples presented deny the Cartesian split that created "brain-in-a-vat," or a Cartesian brain disconnected from the world due to its inability to know anything for certain about objects in the world. "When we say that there is no outside world, this does not mean that we deny its existence, but, on the contrary, that we refuse to grant it the ahistorical, isolated, inhuman, cold, objective existence it was given."[24] In other words, theories create realities based on the "real" as it is understood from each of these worldviews, whether they are liberal, realist, or constructivist, rather than explain objective truths about the outside world.

Another issue at stake is the relationship between knowledge production and its connection to acting in the world or how knowledge acts upon the world. Some in IR take, as the authors' below stress, "a rather naive position" that follows

> an objectivist notion of science whose scholars tend not to expound the problem of contextual influences on knowledge production, but claim a rather pure independence. At least, this is what the underlying imaginary of two different worlds suggests. Knowledge production and theory (IR) here forms one world, different and separated from another, which seems to be composed of pure practice. Analysts struggle to bring together what they once split apart: many of the "increasing relevance" or "bridge-building" recommendations are directed towards reconsidering social influence.[25]

There can be a need, however, to see the "Big Picture" sometimes while knowing full well that "the Big Picture is just that: a picture."[26] After all, this is global politics. To continue this metaphor, what is crucial is how other questions should be raised about the picture, such as "in which movie theatre, in which exhibit gallery is it shown? Through which optics is it projected? To which audience is it addressed?"[27] These questions create a different, and often more difficult, project.

Political Imagination

It is imperative that we, as IR scholars, can see alternatives to the state, for instance, from inside of the very form that gives us the epistemological borders within which we theorize. Our political horizons are defined and defended by the state as the very place to act and exist politically. Paul Feyerabend, in *Against Method*, wrote that to begin to see alternatives to how the world is, "We must invent a new conceptual system that suspends, or clashes with, the most carefully established observational results, confounds the most plausible theoretical principles, and introduces perceptions that cannot form part of the existing perceptual world."[28] In IR, it is nearly impossible to begin a critique against the state because it is the starting point for action and theorizing in the international realm.[29] Further, Feyerabend queries: "Now, how can we possibly examine something we are using all the time? How can we analyse the terms in which we habitually express our most simple and straightforward observations, and reveal their presuppositions? How can we discover the kind of world we presuppose when proceeding as we do?"[30] How indeed?

Following Feyerabend, this book puts forward, in the language of International Relations, a set of external and alternative assumptions "constituting, as it were, an entire alternative world" because "*we need a dream-world in order to discover the features of the real world we think we inhabit* (and which may actually be just another dream world)" (15). The first step in our criticism of familiar concepts and procedures "must therefore be an attempt to break the circle" (15). How do we begin to imagine a different way to "dream" international community?

To begin to respond to this question one must acknowledge, through interpretations of thinkers such as Hobbes, Locke, and Descartes and ending with Kant, that generally IR, as a discipline, holds the world to an idea of nature and the body based on mechanical metaphors with a reliance on mind–world dualism, a transcendental subject that insists morality must take the form of law through "apodictic recognition," and an "understanding of nature through Newtonian laws."[31] In order to open a different realm of possibilities for IR, an exploration into the "Radical Enlightenment," epitomized through the thought of Spinoza, should be undertaken. Especially important for this work, Spinoza in his writings saw, as explained by William Connolly, "morality as law grounded in intellectual love of the complexity of being" rather than recognition based on absolute necessity and truth, dualism replaced with parallelism that acknowledged the relationship between mind and body, and the pursuit of robust democratic pluralism.[32]

Just as there is no claim to objective truth in this book, there can be no facile split between nature and culture, and in many ways the very idea of "nature" may presuppose too much. Nature, writes Donna Haraway, is not a physical place, an "other" that gives beginning or nourishment, or "a text to be to be read in the codes of mathematics and biomedicine." Nature is not a tool or slave to man, but rather is "a topos, a place" that can "order our discourse ... [and] compose our memory."[33]

This stream of political thought allows for a world where "all knowledge is contingent and contextual,"[34] and where we find that there is no radical gap between mind and things. Crucially, this should not create a sense of existential panic that demands a new way of fixing the subject to some sort of new immutable, transcendental law. Rather, it is a matter of concern that takes an interest in emergent forms of life and different ways to organize identities and rights around contingent and complex conditions.[35] These concerns and interests can become a new way to "dream," or organize international relations.

In addition, this way of understanding subjectivity moves beyond a purely phenomenological way of experiencing the world. The human may be in the world as a thinking and feeling subject, but humans are also acted upon by other bodies and forces. The world has an impact on and through the body—focus on affect, or states of relation, forces of encounter, or passages between these, that aid in theorizing bodies by their "potential to reciprocate or co-participate in the passages of affect" rather than entities bounded by skin or some other surface.[36]

Likewise, affect is not limited to the human—as I shall demonstrate through bacteria in and on the human body—but also "forces of temporality, gravity, and magnetism" and matter of all kinds.[37] A body, whether human or nonhuman, can affect or be affected by all other bodies because, as Jane Bennett explains: "*All* forces and flows (materialities) are or can become lively, affective, and signaling. And so an affective, speaking human body is not *radically* different from the affective, signaling nonhumans with which it coexists, hosts, enjoys, serves, consumes, produces, and competes."[38]

Bodily Imagination

The body is in a process of becoming with other bodies in a world that defies our ability to find its final shape or meaning. The body in this "lively" world becomes a body by its interactions with the world and the assemblages of which it is a part. "A body is not defined by the form that determines it nor . . . the functions it fulfills," but as a process of "movement and rest, speed and slowness"[39] and affect, power, and potential. Or in Spinoza's words, *"Bodies are distinguished from one another in terms of motion and rest, quickness and slowness, and not in respect to substance."*[40] And for International Relations in particular, it is important to note that "bodies, as well as objects, take shape through being oriented toward each other, an orientation that may be experienced as the cohabitation or sharing of space,"[41] to discover techniques for caring for these bodies and the worlds they inhabit. New political formations can be built from watching how bodies interact and build their worlds.

To return to the role of metaphor and briefly connect it to the above, the metaphorical language circulating through the body politic aids in understanding the power of bodies to form thought and action. Ways of speaking about the world smuggle in beliefs about the world or specific kinds of ontology along with the words. This work is not about the metaphors per se, but about the nature of the ethics and morals that are supported through these ontological framings of the world. Do we speak of the world instrumentally through metaphors that liken the world to a precision instrument such as a watch or some other mechanism? Is the world a commodity? Or is it a use value that sustains human life? Is it a complicated, highly sophisticated assemblage that keeps us alive? Even if we see the world mechanistically, this will only tell us how the world works or how it was built, rather than how to build something new.

Simply put, if we speak about the world differently, we will treat it differently. This means talking about bodies and worlds differently. Certain kinds of everyday immoral practices would be impossible if the world was spoken of as a site of miraculous contamination that gives life in a cold, empty universe. "The earth," writes Hannah Arendt, "is the very quintessence of the human condition, and earthly nature, for all we know, may be unique in the universe in providing human beings with a habitat in which they can move and breathe without effort and without artifice."[42] Reflecting on this thought alone may drive humans to find ways to better care for the world. It is not about the metaphor, or the rhetoric, but about which ontology, or mode of becoming, supplies the best metaphors for global living, and even thriving. The treatment of language in this analysis is sensitive to relations of power, force, and violence, rather than relations of meaning within symbolic structures.[43] I will introduce different affects, forms of thought, language constructions, and ideas of bodies that foster and guide interest in a world that is irreducible to a simple commodity or mechanical system. As Spinoza wrote in *Ethics* in 1677, "no one has hitherto laid down the limits to the powers of the body, that is, no one has as yet been taught by experience what the body can accomplish."[44] Put simply, no one knows what a body can do. Advances in medical imaging and greater biological understanding of the human body have only increased its wonder and mystery.

The body is never a passive object of abstract moral reasoning stemming from a disembodied subject. It is embedded in the world. "No human is, or ever could be, merely a 'brain-in-a-vat' . . . because it leaves out the critical role of our body-in-interaction-with-our-world that defines human meaning, reference, and truth."[45] This "disembodied subject" is a legacy of Cartesian dualism in the social sciences and accounts, in part, for the body's so-called absence from the social science agenda. It is noted now that the body was never really absent, just "silent and unacknowledged" and a "rediscovery" of this silent body as a legitimate object of study became possible due to many factors.[46]

Contemporary writers stress that the body has become a topic for study because of the modern processes of industrialization, consumer culture, capitalism and its politics, second- and third-wave feminism, changes in medical practice and technology, demographic shifts to an aging population, global pandemics likes AIDS and SARS, advances in cybernetics and virtual reality, and the growth of biocapital, to list but a few. These broad transformations of society, coupled with technological advancements like organ transplantation, xenotransplantation, and medical enhancements to human performance through drugs and surgery, have placed into question the idea of the human body as "natural" and "pure." We can no longer "presuppose a human subject on the lines of the model provided by classical philosophy."[47] In the realm of bioethics, these medical technologies have also shaken the common distinctions heretofore made in relationships, especially the "differentiation between the social interaction of two human agents" and the relations between objects and personal agents.[48]

Roberto Esposito argues there has been a shift from the classical analogy and metaphor between the state and body to "that of an effectual reality: no politics exists other than that of bodies, conducted on bodies, through bodies."[49] Foucault's analysis of how the social body, created in the nineteenth century, became a metaphor that no longer referred to the king's body, but rather resided in the population.[50] "It is the social body that needs to be protected," said Foucault, "in a quasi-medical sense . . . the phenomenon of the social body is the effect not of a consensus but of the materiality of power operating on the very body of individuals."[51] Biopolitical regimes and the body politic are intimately related.

To further unsettle accounts of what it means to be human, the more we know about the body, the less control we seem to have over it as an object. These two paradoxical developments in high modernity have also supplied a background and have been important for the body's visibility as a subject of inquiry. According to Chris Shilling: "We now have the means to exert an unprecedented amount of control over bodies, yet we are also living in an age which has thrown into radical doubt our knowledge of what bodies are and how we should control them."[52] The more we know, the less we seem to understand and control both bodies and the knowledge born from these understandings.

These transformations and advancements have made it of paramount importance to be aware of the role of the body—and life more generally—in politics. Conversely, respect for the lived body is necessary for creating responsive policy. This means that attention must be drawn to "the role played by the body as a visceral antagonist within political encounters" and not just as an object affected by politics.[53] Scholars need a revised view of what "the body" means and this "meaning is a matter of relations and connections grounded in bodily organism-environment coupling, or interaction."[54] Moreover, political theory must attend to the importance of bodies through recognition of their material environment and socioeconomic structures, and further how these, along with other bodies, "dictate where and how they find sustenance, satisfy their desires, or obtain the resources necessary for participating in political life."[55]

Of course, problems do crop up when social action operates with no account of the body.[56] Naturalistic and social constructivist approaches limit the ability to theorize the body. It is important to outline a view of the body as a material phenomenon that is shaped by, as well as shapes, its social environment. I do not ignore the role of social construction, but this analysis works to be sure that social construction and materiality are not over- and underdetermined at the same time. Society and biology are at once real and socially constructed.

The body is also central to our ability to intervene in and have agency in the world. "It is important to recognize that the body is not simply constrained by or invested with social relations, but also actually forms a basis for and possesses productive capacities which *contribute towards* these relations."[57] Bodily emotions, preferences, and actions are a source of social forms and norms. "To rob the body of its own history and characteristics ... is to neglect how our embodied being enables us to remake ourselves by remaking the world around us."[58] And, as I argue throughout, the world and other bodies remake and create our bodies. They cannot be spoken of separately because the body *becomes* a body through its interactions with the world. Study of the actual body and relations of power must be placed before ideology: this is the materialist approach.

Material Imagination

The new materialist approach can be said to concern questions that center the idea of active, agential matter. Appropriately, it is a focus on the "stuff" and things of the world, creating a move toward reality itself as it takes into account the material world. More generally, the nature of reality becomes independent of thought and of humanity. All objects are equally real, if not equally strong. This work is part of a larger project that believes in acknowledging material causality and corporeality and their roles in shaping the world and the politics in the world.

The term *materialism* can be found in a variety of different approaches: Marxism, feminism, queer studies, poststructuralism, and phenomenology. Although it does not deal with the same "stuff" as other materialisms, it does not deny the materialist heritage that came before through Marx, Freud, and Merleau-Ponty. Crucially, Marx emphasized historical over metaphysical materialism, and demonstrated that things that were seen as "natural"—family, market, state—were social constructions that could be recognized as such and, even more crucially, changed. Materialism insists on the power of these terms while at the same time acknowledging the political desire to change these constructed institutions to recognize material contexts. Social justice, normative and ethical theory, and policy creation need to be engaged with these material contexts in order to fully respond to the conditions of this material world.

Differences that have previously set humans apart from other species—self-reflection, reasoning, moral reflection—can be seen as "luck of the draw" on the cosmic level. If humans take a decentralized view of the world and the "stuff" or material of politics seriously, this puts the very categories of politics into question, even the very concept of the political itself.[59] This does not reduce humanity to the rest of nature, but supports an appreciation, as William Connolly asserts, of the flow of agency from "simple natural processes, through higher processes, to human beings and collective social assemblages."[60] These are differences in degree rather than kind.

If politics are viewed as future-oriented, ongoing processes of power relations in an environment of shifting agents, rather than only constitutional or normative, this will necessarily involve a shift in metaphor from a unified and autonomous sovereign body to one based on complex networks of power relations characterized by decentralized, multiple, and dynamic connections. If the human is a hybrid forum composed of nested sets of complex permeable bodies, this leads to a new conception of "*bodies* politic" or a set of evolving and interlocking organic systems within systems. This can transform our imaginary of the world, our place within the world, and the responsibility we have toward the bodies in our bodies, and the bodies in the world. Processes and techniques that address these deep and multiple relationships in the world will lead to changes in ethical and normative frameworks. It is the job of theory to interrogate these conditions. In this book, theory is active and productive, and exceedingly corporeal to mirror the body's vibrancy. The body is a collective; it is an historical artifact constituted by human as well as organic and technological unhuman actors. Actors are entities which do things, have effects, build worlds in concatenation with other unlike actors."[61]

This conversation between IR theory and new materialism allows this work to creatively explore "new corporeal capacities" and reflect "seriously on how we might have been, or could be different than we are."[62] The engagement with materialism is less about defining a theoretical approach than about gesturing toward the problems it can bring to the fore. This approach challenges basic assumptions about modernity, especially those surrounding the role of human agency, subject creation based on this agency, and what this subject's relation with nature and culture entail. Moreover, rethinking these practices and relations leads to the need to address the far-reaching normative and ethical implications that, in turn, become apparent. The addition of the life sciences through metagenomics and immunology offers a view of a complex world of interactions between human bodies and their microbiomes in order to unsettle our place in the great chain of being and see the agentic capability of nonhuman actors. This, in turn, aids in seeing the complexity of global politics in new and productive ways.

To aid in creating a vocabulary that can better describe this plural and relational agency of bodies, this book also draws from the sociology of scientific knowledge (SSK), and science, technology, and society studies (STS). A brief explanation is warranted, insofar as science studies is not generally referenced in IR.[63] In fact, Ole Wæver commented in 1998 that "the relationship between IR and sociology of science is virtually nonexistent," but that it could offer interesting insight into the field and study of IR.[64] I believe that, more than just offering a way to understand IR practice as a "science," science studies can offer IR some resources for affirming global politics as a realm of multiplicity—of actors, bodies, negotiations. Science studies and IR share a common interest in predicting and understanding diverse actors' behavior in complex and complicated environments. Many of the pressing issues in global politics demand policies and actions beyond the scope of the sovereign state: pollution, pandemics, nuclear proliferation, biodiversity protection, and climate change, to name but a few examples. Science and science studies offer concrete ways to address these issues, and IR's relevance as both an academic discourse and a contributor to policy, can be advanced by fostering an engagement with the main ideas of science studies.

STS as a discipline can help in finding terms to try to communicate the imbricated nature of human and nonhuman relations, and moving away from agency as the ability to act according to a humanistic idea of free will. In fact, the ability and impetus to act is complex even among humans. As Latour asks in *Reassembling the Social,* "When we act, who else is acting? How many agents are also present? How come I never do what I want? Why are we all held by forces that are not of our own making?"[65] He also stresses that "Action is not done under the full control of consciousness; action should rather be felt as a node, a knot, and a conglomerate of many surprising sets of agencies that have to be slowly disentangled" (44). To add to this:

> the theory of action itself is different, since we are now interested in mediators making other mediators do things. "Making do" is not the same thing as "causing" or "doing": there exists at the heart of it a duplication, a dislocation, and a translation that modifies at once the whole argument. It was impossible before to connect an actor to what made it act, without being accused of "dominating", "limiting", or "enslaving" it. This is no longer the case. The more attachments it has, the more it exists. And the more mediators there are the better (217).

Contrary to the above explanation, the debate in social sciences based on notions of agency tends to center around how agency is formed by a priori structures that provide constraints and resources.[66] Science studies offers a way, through the anthropological and ethnological study of science practice, to focus "on the complex and controversial nature of what it is for an actor to come into existence" rather than being defined through structures.[67] Latour borrows the word "actant" from semiotics to include nonhumans in this definition, but it must be remembered that the human–nonhuman distinction is used to bypass the subject–object distinction. Graham Harman writes of Latour's take on objects:

> Latour takes apples, vaccines, subway trains, and radio towers seriously as topics of philosophy. Such actors are not mere images hovering before the human mind, not just crusty aggregates atop an objective stratum of real microparticles, and not sterile abstractions imposed on a pre-individual flux or becoming. Instead, actors are autonomous forces to reckon with, unleashed in the world like leprechauns and wolves.[68]

In Latour's words, "Associations of humans and nonhumans refer to a different political regime from the war forced upon us by the distinction between subject and object" or "what the object would look like if it were not engaged in a war to shortcut due political process."[69] The more familiar focus on agency as a thing or a person that acts in a particular way for a particular result is shifted to a complex realm of actants who may not have agency in ways that are familiar to us—such as methane-eating bacteria in the Gulf of Mexico that aided in oil cleanup—but clearly have "trajectories, propensities, or tendencies of their own."[70] These tendencies often have political effects in the human world; the bacteria did not act in response to a directive from the government or BP, but they clearly helped out multiple species in the Gulf with their efforts. This is an ability to act that is independent of consciousness and not entirely human-centered.[71] How can we take these nonhuman bodies seriously in politics?

It is here that science studies can add clarity: instead of starting with a priori entities, science studies, echoing Spinoza, defines the actor by what it *does* rather than what it *is*. This world composed of actants, with one no more real than the other, displaces the subject-oriented idea of agency. Even more important, science studies allows diverse actants to be understood as having the same ontological status and the "isolated Kantian human is no more and no less an actor than are windmills, sunflowers, propane tanks, and Thailand."[72] To return to the above example, it is less important to study the methane-eating bacterium as an actor, and more crucial to understand how it works in the complex ecosystem of the Gulf and how we might create favorable conditions for further or future assistance in environmentally sustainable reclamation.

Because of this diversity of actants, an important dimension of science studies is its interest in creating, or composing, a politics that can take into account more than just humans, because a common world must be composed of all beings and objects that share the earth. This must include acknowledging the connections between humans and nonhumans. This is done, in part, by dismantling the difference between content and context, and subject and object, to understand the many translations that occur between one and the other. The social world is thus "flattened," and the split between nature and culture becomes an unnecessary and harmful distinction for understanding the world. "Translations" replace these rigid dichotomies by focusing on the way in which "actors modify, displace, and translate their various and contradictory interests."[73] These principles do not offer further compromises between the subject/object, micro/macro, nature/culture, local/global, but rather take seriously the "impossibility of staying in one of the two sites for a long period."[74] It is the movement between the two—the translation or the trace of this actant as it creates "controversies"—that allows for a science of the social. These controversies, or where the translations can be traced, are not about fixing actants into categories, but about letting the actants order themselves instead of the analyst. These translations do not make up "society," but rather a "collective" emerges. At the center are the associations between humans and nonhumans, and that creates the possibility of various actants meeting across shared worlds and experiences.

Materializing Words

These translations could also be termed *entanglements*. As discussed earlier, the body and bodily experience are an important source for metaphors, and it is the imbricated nature of the two that is emphasized—where *words and worlds are entangled*—that drives this exploration. Metaphor opens a space to think and understand the world materially as well as linguistically. In other words, I stress the *metaphorogenesis* of concepts in body-based ways of thinking and in the *effectivity of metaphorical framing* in influencing our opinions and judgments about the world. I then extend this idea of metaphor to analyze how this might influence people's thinking and problem solving about the material world.[75]

In this, I can represent ideas as connected to matter and lived bodies without prioritizing or valorizing either as superior. Thus, words and human cognitive processes that make sense of the world by relating abstract concepts to bodily understandings can contribute to shaping the material world. Also, we can understand things and concepts in the world via a range of metaphors and, conversely, they must be understood nonmetaphorically as well. "Part of a concept's structure can be understood metaphorically, using structure imported from another domain, while part may be understood directly, that is, without metaphor."[76] This helps to ameliorate the anthropocentric tendencies of discourse analysis based on metaphors that stress only ideas, or the output of the "brain-in-a-vat" as those that are essential and constitutive of reality. This reliance on language is perhaps one of the most theoretically difficult points when using metaphor, along with an interest in displacing the human's privileged position. After all, it is human language that is communicating these ideas; however, as Jane Bennett writes, "One can point out how dominant notions of human subjectivity and agency are belied by the tangles and aporias into which they enter when the topics are explored in philosophical detail." Bennett urges invoking all manner of nonhuman material and beings entangled with the human in order to "show how human agency is always an assemblage of microbes, animals, plants, metals, chemicals, word-sounds."[77]

Materialism and science studies further work to unsettle a phenomenological human "self" or "subject" that encounters "objects" already existing a priori in the world. Generally, the role of the object is downplayed in cognitive metaphor theory as something static only brought into being by its encounter with human consciousness; it is "the world-for-a-human-consciousness" that focuses on a "narrow world of human intentionality."[78] An expanded understanding of other actants, through the use of science studies and materialist orientations, demonstrate that agency is more complex than simply human language and intention.

The metaphorical approach to the body—as the body politic—can take an account of the power of bodies in international relations even if it does not have the language or ability to do so. "It is obvious that the metaphor of the body underlies a good deal of political discourse" and "words derived from the body translate into words of world politics." The authors of

Metaphorical World Politics list several examples: "Attraction (alliance), balance (balance of power), blockage (blockade), center (center–periphery), collection (coalition), compulsion (compellence), container (containment), contact (contacts)."[79] These often reflect the physical orientations of bodies in space, bodies interacting directly, and bodily associations such as those in the family and daily life.

To demonstrate these points, the focus will shift to the use of metaphor in IR and give examples of slippage between the discourses of medicine and war to demonstrate that the relationship between the disciplines is already present in the literature. This rests upon the larger claim that political science, and International Relations as a subfield of political science, borrows metaphors from the sciences and relies on metaphors and metaphorical frames to communicate many of its central tenets. To introduce this subject more broadly, "metaphors with direct body referents focusing on disease and medicine (germs and microbes) and other living organisms (microbes, rats)" are common in International Relations.[80] Also, Richard Little, in his book on the "balance of power," argues that this general metaphor is "employed ubiquitously to transform the conventional conception of power" and that the balance-of-power concept is essential, central, and privileged in Morgenthau's *Politics among Nations*, Bull's *The Anarchical Society*, Waltz's *Theory of International Politics*, and Mearsheimer's *The Tragedy of Great Power Politics*.[81]

For example, the balance of power is a powerful organizing metaphor in IR: "no other theory has the extended pedigree of the balance of power."[82] It is a "simple but extremely effective metaphor that transforms an agency-based concept of power to . . . a structural concept, where power is a product of the system and the overall distribution of power must be constantly refigured" (13). This metaphor has been "transmuted" into a "long-established myth" used to narrate the survival of Europe as independent states, among other things (13). It is connected to the larger adoption of mechanism models for describing the world around us. Karl Deutsch writes in *Nerves of Government* that the development of clockwork since the thirteenth century created the "mechanism" model that was subsequently

> applied to a description of the stars in the system of Newton; to government in the writing of Machiavelli and Hobbes; to theories of the "balance of power" and "checks and balances" by

Locke, Montesquieu, and the founding fathers of the American Constitution; and to the human body by such eighteenth-century writers as La Mettrie, author of the book *Man a Machine*.[83]

Deutsch adds that this notion of the mechanism was a "metaphysical concept" and that "nothing strictly fulfilling these conditions has ever been found anywhere" (27). The more important point is that this model implied certain assumptions—such as a whole that was equal to the sum of its parts—while excluding others.

To highlight war, metaphor, and their relationship, George Lakoff writes that a common metaphor in which military control by an enemy is seen as a spreading cancer. In this metaphor, military "operations" are seen as hygienic, to "clean out" enemy fortifications. Bombing raids are portrayed as "surgical strikes" to "take out" anything that can serve a military purpose. The metaphor is supported by imagery of shiny metallic instruments of war, especially jets.[84]

In the case of humanitarian intervention, the language that is used often frames the interventionist actions and creates metaphors from the field of medicine. We offer humanitarian "relief," "injections" of foreign aid, and "prescriptions" for ameliorating unrest. These metaphorical frames show that bodies must behave in certain ways, thereby implicitly arguing that certain ideas about healthy bodies and lives must serve as a template for unhealthy ones. Humanitarian intervention is then about healing these "unhealthy" bodies.

In medicine, "germs" are spoken of as "enemies" to good health. We are at "war" against cancer. New research in AIDS helps us to "conquer" the virus. In fact "militaristic language pops up in almost every scientific domain: conservation biology ('invasive species,' 'biosecurity'); global warming ('global war on global warming'); and biomedicine ('killer cells,' 'hitting multiple targets')."[85]

How does this language affect how we speak of enemies and opponents? I offer these examples to demonstrate the violence that can rise with the use of these metaphors:

> *Guardian*'s News Blog, March 21, 2011: "8.13am: Yesterday we heard that supporters of Gaddafi had formed a human shield around his compound in Tripoli, with men, women and children singing songs against the rebel "germs."[86]

Israeli News website: National Union MK Michael Ben-Ari, while speaking at an SOS Israel conference in Jerusalem is quoted as referring to members of leftist organizations as "traitors who must be persecuted at any cost." He called the leftists "germs" and "enemies of Israel" and added "If we'll have to enact a law in the Knesset to eradicate this dangerous enemy, that is what we'll do. Such a germ can destroy Israeli society. This enemy threatens the state's existence."[87]

In the *Sunday Times,* Mark Franchetti interviews a soldier who says, "The Iraqis are sick people and we are the chemotherapy," said Corporal Ryan Dupre. "I am starting to hate this country. Wait till I get hold of a friggin' Iraqi. No, I won't get hold of one. I'll just kill him."[88]

What does this circulation of medical and military metaphors do? A crucial point is that "analogies and metaphors are the often hidden devices that serve to rationalize choices of conceptual variants and to render plausible the narrative in which our political concepts are embedded."[89] In other words, by perceiving war and intervention as a "treatment" for sickness, we are blinded to other responses that may address the situation more effectively. These "hidden devices" can also invite and justify violent responses as shown above in the name of protecting health. Which metaphorical constructions make this circulation a possibility? What makes all these ways of speaking about intervention, states, and war, so compelling, so facile? In part, it is due to our guiding metaphors of the state and the body.

Drawing from Lakoff, the state is often conceptualized as a person or body that engages in social relations and has a distinct personality and disposition; it is understood by giving the state characteristics and motivations we understand through personification metaphors. It can be warlike or peace loving. Its territory is its home, and neighbors and the "family" of nations surround its home. A state uses its strength and possesses health as a human body does, and "a serious threat to economic health can thus be seen as a death threat."[90] The logic of these metaphors allows the state to do whatever is necessary to stay strong and healthy, which can include maximizing wealth and military might for the well-being of the state, regardless of the effects this may have on the population.

Reinforcing this personification metaphor is the metaphor of the container used to understand both the body and territory. "Each of us is a container," George Lakoff and Mark Johnson write, "with a bounding surface and an in–out orientation," and they continue, "there are few human instincts more basic than territoriality."[91] We feel pain "in" our shoulder; we move from one room into another; we hike into and out of the woods, even if the line between trees and meadow is fuzzy. For the state, Anthony Giddens's term puts this belief into clearest relief: "A nation-state is, therefore, a bordered power-container," and the state is the "pre-eminent power container of the modern era."[92] Georges Canguilhelm, tracing the sovereign subject from the Renaissance to Enlightenment, writes that the idea of the body as an "ostensibly autonomous being whose autonomy is manifest not in the exercise of individual will, agency and choice, but in a bounded corporeality that is assumed to end with the skin" is a powerful trope.[93]

This so-called instinct of territoriality and autonomy is applied to both the borders of the state and the body. To return to intervention and the use of biomedical language, it becomes clear that the logic is consistent with seeing the state as a body whose health is threatened by "rebels" and "leftists" as it would be by germs and viruses. Invasive techniques are used to "fight" cancer, and therefore enemies can be treated as tumors would be. On a larger scale, war becomes a fight between two people and "a just war is thus a form of combat for the purpose of settling moral accounts."[94]

Metaphorogenesis

I would like to return to the relationship between the source of the metaphor and its target. Recall that metaphors are based on source domains speaking to target domains, or referents, and this is what makes the metaphor strong and effective. If the argument is not whether we rely on metaphor to understand the world, but instead that the world is shaped through these metaphorical constructs, then this relationship can be *transformative* as well as descriptive. This is the effectivity of metaphor.

Richard Little, when writing of the balance of power in IR as a metaphor, asserts that "when the metaphorical status of a concept is taken seriously, then the effect is dramatic because the source of the metaphor (balance) has the ability to transform the accepted meaning of the target of the concept (power)."[95] Metaphors "focus the mind on certain previously unseen aspects of any area of inquiry," and the discovery, creation, or emergence of new metaphorical framings can open up political possibilities for International Relations.[96]

Although IR scholars can be quick to point out when foreign policy officials use and misuse metaphor, they, as Michael Marks writes in *Metaphors in International Relations*, "have been less inclined to turn the lens on themselves to interrogate the metaphors they use to analyze international affairs."[97] He further stresses that metaphors are integral elements in conceptualizing and creating knowledge in IR, and that paradigms used to explain IR are "built on metaphorical imagery that provides the very theoretical propositions these paradigms use to hypothesize and make predictions about international affairs" (4). He gives the example of the concept of "structures" as they are understood in international relations and this metaphor's role in explaining what governs international relations. Initially it was a way to explain certain metaphorical constraints on states and other international actors, such as "anarchy," but now "it is used literally as a descriptor for that which defines world affairs" (5). The question of intention is less important than the realization that "the language that . . . is chosen is both inevitably metaphorical and influences the way concepts in international relations theory are framed" (5–6).

The affectivity of metaphor also works to stymie, reify, hypostatize, and obscure issues, concepts, and factors in the target domain. Or put differently, "as repositories of cultural understandings, metaphors are some of the principal tools with which dominant ideologies and prejudices are represented and reinforced."[98] The war metaphor and its focus on the attack and defense role of the immune system, or the recognition and elimination of the nonself, obfuscates the equally important examples of cooperation, altruism, and coevolution of different species and their relation to the human immune system. I will address this problem explicitly in chapter 3.

To add to this, many political practices that use metaphor work to reduce complexity and tensions down to a choice between two supposed opposites.[99] This can have the effect of dangerously simplifying the responses necessary to respond to dangers or to the challenges presented. R. B. J. Walker gives the example of the currently popular use of metaphors relating to the balance between liberty and security inside the state, and how this practice has come to legitimate many forms of violence (246). This metaphor works to define liberty and security as similar values insofar as it seems sensible and rational that these two should balance; one must give up liberty for the sake of security. This enables specific policy responses that hope to restore a simple balance between two terms. This tension between these two terms lies at the heart of liberal democratic societies, and the relationship between liberty and balance may not necessarily be expressed perspicuously by the concept of balance. It allows for too many "slippery possibilities for propaganda" (246), and has led to many attacks on personal liberty and the rule of law. The debate over security measures at airports comes immediately to mind. A more serious example lies in the suspension of habeas corpus rights for prisoners in custody for alleged terrorist activity against the United States and its allies. This metaphor also shifts the burden to civil society by forcing the people to choose between two values, thereby lessening the responsibility of sovereign authority for exercising this power. This leads to illiberal and antidemocratic policies made in the assumed interest of popular desire for safety over liberty (247). Further, and perhaps more significantly, Walker asserts that this "need to rebalance liberty and security now sustain attempts to switch back to a more 'natural' account of political identity" based on biometrics, and the "real flesh-and-blood being of each person" rather than "political identities guaranteed by birth certificates and legal signatures with the jurisdiction of sovereign law" (248). This is a move that threatens to undermine our very idea of a modern citizen that is based on liberties defined by something other than a connection to an essentialized natural world beyond political authority (248). In short, metaphorical effectivity can have undesired effects or support conservative policies.

In Latourian language, metaphors are actants in the political realm: forming relations, developing meanings, and shedding light on the discursive and material foundations of the political process. They can be vehicles through which power operates; therefore an awareness of the use and deployment of metaphors is crucial to challenging or transforming these relations of power.[100] Metaphor can also be a way of illustrating how we deal with change and how we resist change, and not just an exercise in "bad metaphysics."[101]

The pervasiveness of metaphor in everyday communications should also be noted. A study conducted on the effect of metaphorical framing and social policy in 2011 and 2015 found:

> Metaphor is clearly not just an ornamental flourish, but a fundamental part of the language system. This is particularly true in discussions of social policy, where it often seems impossible to "literally" discuss immigration, the economy, or crime. If metaphors routinely influence how we make inferences and gather information about the social problems that confront us, then the metaphors in our linguistic system may be offering a unique window onto how we construct knowledge and reason about complex issues.[102]

The researchers in this study stressed that any complex situation requires metaphors in order to convey understanding and meaning. "Metaphors aren't just used for flowery speech. They shape the conversation for things we're trying to explain and figure out. And they have consequences for determining what we decide is the right approach to solving problems."[103]

New metaphors based on the human experience in all its variations and framings can open new paths for demystifying International Relations and clarify how politics and society work. Inasmuch as all human interaction is based on the language of metaphors, the more we put metaphorical imagery to work, the more we can gain comprehension of what human interaction is about, on all levels of existence.[104] Awareness of a metaphor leads to an ability to change perceptions and actions based on that metaphor and illustrates how change is made possible or not.

Importantly, *The Microbial State* focuses on the "ontologically creative aspect of metaphor," or the "creative-productive function that they [metaphors] have in politics,"[105] and in International Relations. This creative function is connected to the material world as well as the linguistic one, by situating the body within the cognitive theories of metaphor. We understand the world, in part, through metaphors created through the experiences of our bodies. We embody and create meaning through metaphor. It is not just a corruption of speech, a substitute for a literal form, an utterance or a lens through which an object is viewed.[106] This body is not an a priori subject. In this, I want to bring the cognitive metaphor theorists a step further into the world than they may have originally intended. The body is clearly not just biological or social, but structured by relations of power and created through its interaction with other bodies.

2

Lively Subjects, Bodies Politic

> *I have all lives, all effects, all causes, all germes, invisibly hidden in myself,*
> *This is the earth's word—the round and compact earth's,*
> *I and the truth are one, we are curiously welded*
>
> —Walt Whitman, *Leaves of Grass*

Meet *Bacteroides thetaiotaomicron* (Figure 2). This rod-shaped and gram-positive bacteria is found in the human intestine, where it aids in breaking down polysaccharides that the human body would not otherwise be able to process. Approximately ten trillion bacteroides are packed into the gut, and about 15–20 percent of our daily caloric intake is accomplished through bacteria like this one "fermenting foods like carbohydrates and producing various acidic waste products."[1] These bacteria and other gut microbes function as "key interfaces" between our bodies and the environment; "they defend us from disease-causing pathogens that cause diarrhea, and they detoxify potentially harmful chemicals that we ingest."[2] Some studies suggest that gut microbiota can alleviate autoimmune disorders such as allergies and irritable bowel syndrome.[3] *Bacteroides* also produce heat in the same way fermentation creates heat, and this heat is used to regulate human body temperature.[4]

There are ten million species of bacteria and over five thousand species of viruses. These microbes are too small to see with the naked eye, but make up half the biomass on the planet and have exerted influence on the course of human evolution and history. Humans also have positive and negative long-term relationships with microbes. Well before we knew of the microprocesses involved, humans used bacteria to ferment beverages and cheeses and to preserve olives. Plagues and diseases affected bodies, populations, and events across continents, changing the course of human history.[5]

Yet, with the exception of pathogenic viruses and bacteria, microbes rarely register as objects important to politics. This is partly a matter of scale; humans are the largest organisms, and microbes are the smallest. Bacteria belong to a universe of single-celled creatures too small, with rare exceptions, to be seen by the unaided eye, but they exist everywhere on

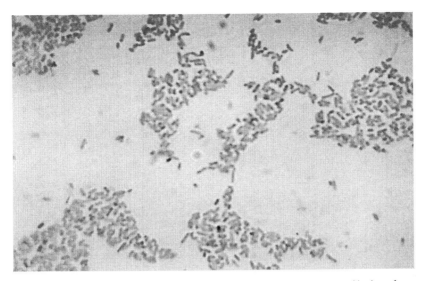

Figure 2. This photomicrograph reveals the bacterium Bacteroides thetaiotaomicron *grown on blood agar for forty-eight hours. Content provider: CDC/Dr. V. R. Dowell Jr.*

Earth. Being small, simple, and many confers on bacteria advantages that allow them not only to survive but also to affect every mechanism by which the planet works. In *Living at Micro Scale*, David Dusenbery asserts that it is difficult for humans to conceive of the physical situation of bacteria. "Imagine," he writes, "stepping off a curb and having to wait days for your foot to reach the ground."[6] Practically, this means humans have much more experience and relevant knowledge toward understanding and designing for larger things rather than smaller things and think more readily at the macro level.

Microbial Relations

Stewart Brand, an American writer and ecological activist, asks: "When confronting a difficult problem we might fruitfully ask, 'What would a microbe do?'"[7] What follows takes this provocation seriously. The topic of this chapter is metagenomics, or the study of collective genomes directly from the organism's environment, as a way to envision political community as an assemblage of multispecies groupings and highlighting analogies born from the relations between humans and microbiotic communities that fosters and creates a sense of deep plurality within communities. "The New Science of Metagenomics: Revealing the Secrets of Our Microbial Planet," is an exemplary text to introduce both the details and definition of what metagenomics is, as a science and as a way to demonstrate how biological discoveries have begun to shape opinion about how humans see themselves and their relation to the larger world.

Reactions in the popular press to the discoveries of the Human Microbiome Project—a study of the human bacterial communities undertaken with metagenomic sequencing techniques—point toward a shift in the metaphorical references of the body. Human bodies are "planets," "superorganisms," and "ecosystems" with microbes as "helpers" and "agents" rather than "dangers," "enemies," and "pests." To bring metaphoricity to the fore, I call bodies, as they are understood by their microbial relationships, *lively vessels*.

This lively vessel is used to destabilize the perceived autonomy and individuality of the human and, in the following chapter, to complicate the "state as person" metaphor relied upon in International Relations (IR). This is done to begin to think about the lessons we can take from *Bacteroides thetaiotaomicron,* and how our understanding of the body politic could be different if the human microbiome is used as perspective. How does this perspective change the vocabulary of International Relations? How do the microrelationships within the human body speak to the macrorelationships in international relations?

Metagenomics is the study of the microbial world—including the above bacteroides—from an aggregate level, "transcending the individual organism to focus on the genes in the community and how genes might influence each other's activities in serving collective functions."[8] In the above report the authors write that metagenomics is a new science that combines genomics, bioinformatics, and systems biology to study the genomes of many organisms simultaneously. Most microbes have essentially been invisible insofar as the majority cannot be grown in a laboratory or studied with the methods of microbiology.[9] Metagenomics is operationally novel and involves more than just listing and cataloguing sequences and microbial diversity.

> The inventories also give unique insights into microbial community structure and biogeography. They enable subtle understandings of

ecophysiological characteristics of communities, in which adaptations to different environmental gradients result in different metabolic and morphological strategies (e.g.: capacities for movement) that spread vertically and horizontally through community members.[10]

Previously, almost all knowledge about microbes was found in the laboratory "attained in unusual and unnatural circumstances of growing them optimally in artificial media in pure culture without ecological context."[11] This encouraged the study of so-called lab weeds, or only microbes that did well in aseptic environments. The ability to study microbes in their natural environment has revealed that we can better understand a microbe by knowing where it fits within its community and adjacent communities rather than isolated in the lab. The previous mechanistic framework of the "abstraction of an ontology of fixed entities" may need to give way to the idea that "life is in fact a hierarchy of processes (e.g.: metabolic, developmental, ecological, evolutionary) and that any abstraction of an ontology of fixed entities must do some violence to this dynamic reality."[12]

This ability to study entire communities of bacteria opens up new applications and areas of study within biology, medicine, ecology, and biotechnology. "Traditional microbiological approaches have already shown how usual microbes can be; the new approach of metagenomics will greatly extend scientists' ability to discover and benefit from microbial capabilities."[13] Basic ideas in biology will need to be rethought and studied anew: organism-constituted hierarchy, teleological constructions of evolution running from unicellular (simple) to advanced (multicellular), and the mechanistic models of fixed entities.[14]

The Human Microbiome Project (HMP) is one of several international projects undertaking studies to use metagenomic analysis to study human health by analyzing the human microbiome. It was initiated in 2008 to undertake a five-year project to build and create community resources in the emerging field. The microbiome consists of all the microorganisms that live in or on the human body, and, notably, the microbiome is an ecosystem in which all the various members maintain balance and equilibrium. These communities inhabit the mouth, skin, guts, and respiratory and reproductive tracts of human bodies. Some of these microorganisms cause illnesses, but many are necessary for good health.

Importantly, "these functions are conducted within complex communities—intricate, balanced, and integrated entities that adapt swiftly and flexibly to environmental change."[15]

In May 2010, the HMP published an analysis of 78 genomes in the human body, and reported that, genetically, microorganisms outnumber human cells in the body by a ratio of ten to one. A survey of news, magazines, and journal articles about the human microbiome reveals a wealth of colorful statements about this discovery (emphasis added):

> We think that there are 10 times more microbial cells on and in our bodies than there are human cells. *That means that we're 90 percent microbial and 10 percent human.* There's also an estimated 100 times more microbial genes than the genes in our human genome. So we're really a compendium [and] an amalgamation of human and microbial parts.[16]

> *Outnumbering our human cells by about 10 to one*, the many minuscule microbes that live in and on our bodies are a big part of crucial everyday functions. . . . But as scientists use genetics to uncover what microbes are actually present and what they're doing in there, they are discovering that the bugs play an even larger role in human health than previously suspected—and *perhaps at times exerting more influence than human genes themselves.*[17]

> We're each like a *superorganism*—a unified alliance between the genes of several different species, only one of which is human.[18]

> We are not just the expression of an individual human genome. We are . . . *a genetic landscape.*[19]

> We human beings may think of ourselves as a highly evolved species of conscious individuals, *but we are all far less human than most of us appreciate.* . . . We are only beginning to understand the sort of impact our bacterial passengers have on our daily lives.[20]

> *Maybe we aren't quite as human as we thought.* . . . But however you feel about the "trillions of friends" that make up your microbiome, one thing is sure. You're never alone.[21]

These quotes demonstrate how discoveries, like those exemplified in the HMP, have affected popular imaginings of what it means to be human and the intra- and interconnections with nonhuman objects. Taken together, ideas about the relationships between humans and their microbial communities have shifted from the definitional "book of life" to the more expansive "communities of life." Microbes are imagined as friends and as boons to humans, rather than enemies. Even more important in this article is the authors' finding in their discourse analysis of news and science publications that "humans are no longer conceptualized as the pinnacle of evolution, standing apart from microbes seen as populating the bottom of the evolutionary tree: we are all human–microbe hybrids" and new ideas of life itself: "life has to be seen in a much less deterministic and much more fluid and flexible way."[22]

The architects of the HMP and the scientists involved in this research are also aware of the implications of changing metaphors and perceptions. A program funded by the National Institutes of Health (NIH) titled "Human Microbiome Research and the Social Fabric,"[23] gathered thirty interdisciplinary participants to evaluate current ethical, legal, and social issues surrounding microbiome research. The HMP is heralded as an opportunity to reexamine a host of issues: biomedical ethics, public health, property law and the human body, and privacy.

The NIH discusses the HMP's effect on self-identity specifically. They report that the HMP is likely both "to transform how we think of the microbes on and in our body, from enemies that must be eradicated to entities that are important in maintaining health" and "change our concept of the human organism and affect the distinction between us and our environment." They recommend that care should be taken on how HMP education is approached with the public and that "researchers and clinicians need to be mindful when developing language to describe microbial inhabitants."

New understandings of genomes, organisms, and species have led to changes in how biological functions and systems are understood. The Committee on Metagenomics stresses this interest in new ideas. They herald a "grandmother" of a paradigm shift in biology: "It will refocus us one level higher in the biological hierarchy (molecules, cells, organisms, species, populations, communities, the biosphere). It will shift the emphasis from individuals to interactions, from parts to processes."[24] As Eric Juengst points out, "the proponents of the Human Microbiome Project have much grander ambitions: they seek to spark nothing short of Kuhnian scientific revolution in our understanding of human biology."[25] Kuhn writes that "extraordinary episodes" in the practice of science are named scientific revolutions: "They are the tradition-shattering complements to the tradition-bound activity of normal science."[26] The HMP and metagenomics supplement and work along with other dynamic systems approaches in social science, theory, and the humanities, including complexity theory, object-oriented-ontology approaches, new materialism, and network theory.

Conceptually, metagenomics implies that the human body as a communal gene pool, and its "self-extending symbioses," are "highly adaptive and robust against environmental perturbation" and "dynamic, self-sustaining and self-repairing processes."[27] Metaphorical framings built from these understandings of microbial communities aid in bringing system-based understandings of complex processes to the international realm. Many problems that the state in its current form has been unable to address—warming oceans, pandemics, climate change, flows of immigrants and migrants—may be easier to address.

The language and affect involved in explaining the complexity of metagenomic discoveries to the public and the scientific community's further interest in "the social fabric" and microbes' relation to humans may demonstrate that there is a space for creative revision that these metaphors can open for shaping politics with objects as actants. By transgressing the boundaries of our autonomous selves and acknowledging the symbiont colonies in our bodies, we can begin to see the ethical and social implications made possible by relations with diverse objects. Put differently, the HMP and metagenomics can provide new concepts that are better equipped to supply IR with different forms of practice for sustainable, ethical, global living with one another, and with other life forms, as a *bodies politic*. These practices can operate at the micro- and macro levels through the way in which IR understands the body and how these bodies exist together in biological *and* political communities.

Lively Vessels

To return to the main form of international political community, the state, it is important to stress that states are increasingly understood in the globalization and modernization literature as porous, but that this framing often leads to a discussion of limits to the efficacy of states in the face of open borders rather than to new possibilities for action that this porosity may enable. Territory and bodily integrity are protected as natural and desired states of being. This chapter argues that the opposite is the case. Borders can appear as places of healthy invigoration both politically and socially. This is opposed to the inside/outside logic of containers that inscribes these areas as danger zones of Otherness and threat.

The body, with its ever-shifting communities of microorganisms that contribute to its survival, becomes a fecund and material model for politics, and for the body politic writ large, that is, for the state: this is the lively vessel. The body as a lively vessel to complicate the referent in the body politic metaphor—or the *body* in the body politic—is offered without suggesting that metaphors are dispensable and replaceable with clear and unambiguous concepts. The conceit of the lively vessel focuses attention on the liminal spaces in the state, such as the border or zones of political debate contrary to the status quo.

This necessitates a return to the ideas of the body and how these conceptions—both metaphorical and material—create different implications for the subject of politics. The human body is colloquially understood metaphorically as a container bounded by the skin. "Each of us is a container with a bounding surface and an in-out orientation" and we project this onto other physical objects.[28] In other words, a common way in which we experience the world is through our body as bounded by skin as a kind of wall, and we interact with others across this wall.[29] This ontological metaphor based on understanding the body as a discrete, bounded territory not only helps us, as stated above, understand our world in terms of objects or substances, but also "allows us to pick out parts of our experience and treat them as discrete entities or substances of a uniform kind. . . . And such defining of a territory, putting a boundary around it, is an act of quantification."[30] We can then "refer to them, categorize them, group them, and quantify them—and, by this means, reason about them" (25). In fact, many argue that "there are few human instincts more basic than territoriality" (30).

Our body as a container certainly seems true for much of our interaction with the world. We understand ourselves as separate beings and recognize others as such. In romantic moments, we do not actually melt into the arms of our lovers and, at the simplest level, every organism needs a boundary between it and the outside. All of the molecules would simply "wander away" and the organism "would no longer exist."[31] But even this boundary is not as clear as it would seem. "At the same time though, life needs a connection to the outside world. An organism must draw in new raw materials to grow, and it must flush out its poisonous waste. If it can't, it becomes a coffin," but it also must avoid "becoming a blur" (20–21). Our cells carefully regulate this flow and our skin is a barrier, but it also "must be pierced with holes—for sweat glands, ear canals, and so on" (21), to both maintain its integrity and interaction with other bodies.

This conception of a container is also occasionally disrupted by a case of the common cold or food poisoning from the mayonnaise at the salad bar. We are reminded that our bodies are open to the world in ways we often do not understand or see. Viruses and bacteria "invade" our bodies, leaving us sick, but what does it mean to be sick to an organism that has integrated various others into what Margulis and Sagan call a "loose committee" and a "walking assemblage" rather than an individual?[32] They further ask: What if the pathogens that "attack" us are a "part of the committee that makes each of us up to begin with?" (19). This becomes more about various ecological relationships and balance, rather than resistance to invasion (19).

The limits of the container as a metaphor of the body are revealed when this perspective is emphasized. While we are "contained" in a particular form, our bodies also experience "traffic" in and out. The contents of this container are not passive and inert, but busy with their own processes and cycles. As will become clear in the next chapter, the body does not contain these forces and processes, "but rather is touched by and opened up to some of the possibilities of becoming otherwise."[33] There are also, as Katherine Hayles writes, "ambiguities of the plural."[34] Just as sickness becomes a more complex matter, so too, do our notions of self become complicated by these new models of subjectivity. If human life is seen as embedded in complex, material worlds, then,

> Each person who thinks herself this way begins to envision herself or himself as a posthuman collectivity, an "I" transformed into the "we" of autonomous agents operating together to make a self. The infectious power of this way of thinking gives "we" a performative dimension.[35]

The body is not something that "is" (being), but rather is understood as embodying particular qualities and beings that are expressed differently depending on what is happening inside, through, and outside of it in both time and space (becoming). If, as Gilles Deleuze posits, politics is a promise of—or an enacting of—the future, then this body becomes more capable of engaging in these politics than a static, predefined one.[36] This materialist approach is a way to acknowledge the specificity of the human body while simultaneously illuminating that difference is not found in an already defined being, but in heterogeneous forces and intensities at play in and around the human body.[37] These forces "mold our live relations" with the world and are reflected back into the body in a multiplicity of ways (xxiii).

I use this idea of the lively vessel to get at the involvement and fecundity of the human as "superorganism." While we do experience our bodies as containers at some levels, these containers are busy, robust, entangled, and teeming with connections and relationships—much more so than previously imagined. These bodies are not shells with skin stretched across lonely organs: these are nested sets of permeable bodies. We are "multi-componented beings."[38] Put differently,

> the body is not bounded by the skin, where we understand the skin to be a kind of container for the self, but rather our bodies always extend and connect to other bodies, human and non-human, to practices, techniques, technologies and objects which produce different kinds of bodies and different ways, arguably, of enacting what it means to be human.[39]

Or as science writer Robert Krulwich writes:

> While tentative, this science is telling us that each of us is a package composed of Mommy's genes, Daddy's genes and bacterial genes that have moved in and stayed. We have a genome and we also have a "microbiome." And the microbiome is a whole new window into who we are and are going to be.[40]

Recall from the previous chapter that bodies are created through a process of becoming with their environment. In less theoretical terms, this is the evolutionary theory that we have coevolved with a community of bacterial, archeal, and eukaryotic bodies to become what we have named "human." Following Margulis and Sagan's book *Acquiring Genomes*, "We could learn from their miniature technologies and incredible range of symbiotic associations—if we wished."[41]

> Such "self-extending symbioses"... are the evolutionary norm because they are highly adaptive and robust against environmental perturbation. All of these insights, whether delivered by metagenomics or more general molecular microbiology, encourage us to see any animal as a composite of all three domains (bacteria, archaea and eukaryotes) and our own primary genome as a metagenome of microbial and human DNA.... It is even suggested that humans and other animals could be regarded as "advanced fermenters", the main role of which is to house, nourish and assist the reproduction of an enormous array of microbes.[42]

The body is explained and created, in part, through symbiosis. In the evolutionary process, "the establishment of a new form from such a symbiosis is known as symbiogenesis."[43] Symbiogenesis is the horizontal transfer of genetic traits rather than the vertical transfer of traits through neo-Darwinian evolutionary theory; it focuses on the long-term interactions and physical associations between individuals of different species. A classic example of this is the lichen: it is neither green alga nor a fungus, but a coevolution of at least three species. Cooperative adaptations increase the odds of survival. Therefore, evolution is based less on competition within populations and more on organisms working together to create more stability for all involved. Evolution is not solely based on Darwinian competition, and "such evolution requires new thought processes" (15).

Margulis and Sagan write that "New metaphors to reflect on permanent associations are needed." This way of understanding relationships cannot be reduced to a paradigm of "cooperation" over "competition" (19). "Symbiosis, merger, body fusion, and the like cannot be reduced" to the anthropocentric terms used previously (15). They propose "a new search in the social sciences for terms to replace the old, tired social Darwinist metaphors" (15). These phrases "paint with too broad a brush" and they often "miss the complex interactions of live beings, organisms who cohabit" (16).

I am not arguing, and Margulis and Sagan are also quite clear on this point, that society would be better served by symbiotic metaphors rather than competitive ones (17). The range of interactions—both in our bodies and in society—is too complex to simply exchange competitive metaphors for those based on cooperation. For example, parasitic relationships can exist on a continuum from beneficial to deadly to the host. Parasites will take center stage in the next chapter as a way to further explicate this point.

These relationships have to be understood contextually with a focus on the processes at stake. There must be an acknowledgment of these "close associations" formed between humans and their symbionts. There are not "costs" or "benefits" or "competition" or "cooperation," and "struggle" is only the bald fact of trying to reach biotic potential. It is "simply a fact of life," but it is also only part of the story of evolution.

> This may suggest that the organization of life is determined not by competition but by the ability to cooperate (albeit competitively). To put another way, although there is little doubt that competition and selection have been essential to the evolutionary process, it may be that the main respect in which organisms (or cells) have competed has been with respect to their ability to cooperate in complex single- or multi-species communities. If this is the case, then we might expect—contrary to the orthodox evolutionary view that altruism is exceptional and requires special explanation—that the norm among organisms is a disposition to act for the benefit of other organisms or cells. Darwin was right about the general picture of the organic environment being fundamental to the determination of fitness, but his account of the relationships between organisms in those environments needs to be supplemented.[44]

This can, in part, be done by looking at the human idea of an individual as a social phenomenon and comparing it with the idea of the organism in the biological sciences. This is rethinking how we see "individuals," or in Deleuzian terms, how we "dividuate,"[45] differently.

Metagenomics asks, "what is an organism?" once one takes into account the trillions of other bodies in the human that are not human and can speak to how we can understand processes that "dividuate." This complexity and diversity in the human body supports an enriched understanding of how the "human" has been understood as connected to other human bodies, species, and its environment. "The basis for these ambitious claims lies in the radical way in which metagenomics re-conceives biological individuality. Traditionally, the individual mammalian organism—the horse or the human—has been the paradigm for both folk and scientific concepts of biological individuality."[46] The human is recast in light of its "intrinsically polluted nature" and that "the environment is actually inside human bodies and minds," and this can be understood as a novel way to imagine connections between nature and culture.[47] The way we interact with other humans, the microbiotic communities of the oceans, and other animals, for instance, becomes a different endeavor both ethically and practically.[48] A biologist muses about what could be learned from the study of bacteria for the scientific community:

> I learned about microbial communities and what they can teach us about thriving within constraints. Numerically and by biomass, bacteria are the most successful organisms on Earth. Much of this success is due to their small size and relative simplicity, which allows for fast reproduction and correspondingly rapid evolution. But the price of small size and rapid growth is having a small genome, which constrains the diversity of metabolic functions that a single microbe can have. Thus, bacteria tend to be specialized for using just a few substrates. So how can simple bacteria thrive in a complex environment? By cooperating—a cooperation driven by need. The composition of microbial communities is driven by both the interaction of different species and external environmental factors that determine resource availability. Scientists want to learn the rules governing these complex relationships so they can reengineer bacterial communities for the production of useful substances, or for bioremediation. Perhaps as we learn the optimal strategies that microbial communities use to work together effectively, we will gain insights into how we can better work together as a community of scientists.[49]

Metagenomics and symbiosis can give insight into how bodies might be understood as highly sophisticated assemblages that require complex relationships with multiple other things on this planet rather than hard-shelled, bounded-by-the-skin individuals that collide and bounce off each other in empty space. The ontology that is smuggled in the metaphor of the lively vessel requires "careful forbearance" and a delicate touch such as might be needed with "unruly relatives to whom you are inextricably bound and with whom you will engage over a lifetime, like it or not" rather than strangers and aliens kept at bay "outside" our borders.[50] Being human may be less about our individuality and uniqueness than it is about our *inter- and intradependence*. In other words, we may need to refuse who we think ourselves to be rather than discover who we are.[51]

View from the Inside

With the aid of the human microbiome, "the completely self-contained 'individual' as a myth can be replaced with a more flexible definition" that takes into account the role of community and environment.[52] "We must begin to think of organisms as communities, as collectives. And communities are ecological entities" (20). This is all to say that the concept of the individual—or the autonomous rational self in politics—has much in common with the "organism" in biology. Like other metaphors borrowed from the sciences, there is a certain amount of amnesia regarding its origin. Linking the metaphorical/political and the scientific/biological will destabilize the myth of the individual in politics just as the idea of the organism is being rethought in biology.

The human has been defined historically, culturally, and biologically as an individual, or an organism, but data emerging from metagenomics studies give evidence that "each organism does not have exactly one genome" and that "this idea of 'one genome, one organism' starts to look like a poorly grounded dogma."[53] The idea that life is organized around the pivotal unit of the individual organism, which is traditionally conceived of as an autonomous cell or a group of coordinated cells with the same genome, is challenged by the idea that "life is not composed in a machine-like way out of unchanging individual constituents. Genomes, cells, organisms, lineages are all assemblages of constantly changing entities in constant flux."[54] As Juengst says, "Biologically, at least, we are not intrinsically individuals, but collective super-organisms, assimilating multiple species and millions of individual organisms."[55]

Recall, also, that understanding metaphors is contingent upon a concomitant understanding of the culture from which they were created and that these understandings can highlight and hide certain aspects of the metaphor. "When metaphors and the stereotypes they promote go unaddressed they tend to prevail," and these stereotypes are then manipulated for political reasons.⁵⁶ We are already aware that society is a framework of social commitments, but what needs to be placed in this framework in order to support sustainable, ethical interactions?

By way of example, the "individual" places the separateness of the human being as its most important trait. In Kantian and humanist terms, an individual is always an end, not a means; therefore, the will of the individual cannot be sacrificed without consent. In rational choice theory, individual choice leads the individual to maximize preferences without questioning their ethicality or referencing a wider world of other actors. The preferences are better served through accumulating wealth and power. In so-called free market, or laissez-faire capitalism, this has led to a grid of social commitments that place the safeguarding of individual freedom at the center while naming other collective duties "externalities." This places an emphasis on increasing individual wealth and making "choice" more important than emphasizing the "view that individual freedom is of value only when community guided."⁵⁷ Freedom may be another concept that begs for redefinition in the twenty-first century. With so many earth systems on the verge of collapse through human terraforming and resource extraction, and the explicit linking of capitalism to ecological decline, it behooves the humanities and the social sciences to turn a critical eye to all concepts that found final form in the nineteenth and twentieth centuries. Humans can no longer live in their "tiny skull-sized kingdoms" and expect to survive into the next century.⁵⁸

To return to the source of the metaphor, it becomes clear that organisms (or individuals) are abstractions from larger collective entities that regulate the actions of the organisms, and that genomics-based strategies have fewer preconceptions about what makes a single organism.⁵⁹ This brings the biological body more in line with the materialist and radical empiricist's idea of the body as formed through the environment in which it is embedded.⁶⁰ This body is never an a priori one, but evaluated as "expressions of certain movements, sensations, and interactions with their environments" rather than its morphology, or the form that the body takes.⁶¹ As Donna Haraway writes, organisms are made as objects of knowledge, rather than born. "Organisms are biological embodiments; as natural-technical entities, they are not pre-existing plants, animals, protistes, etc., with boundaries already established and awaiting the right kind of instrument to note them correctly. Organisms emerge from a discursive process."⁶² She further explicates that "'objects' like bodies do not pre-exist as such" (298), but that these organismal boundaries come into being among humans, nonhumans, machines, and other instruments.

Creative Revision

Metaphors can create reality through their ontological constructions. A metaphor may be a guide for future action. Such actions will, of course, fit the metaphor. This, in turn, reinforces the power of the metaphor to make experience coherent. In this sense metaphors can be self-fulfilling prophecies. The metaphors involved in explaining the complexity of metagenomics aid in opening a space of creative political revisions.

Groups of metaphorical framing are emerging to explain the findings from the Human Microbiome Project. Three metaphors are widely used by the scientists promoting this initiative:

1. that the human *genome* should be understood as part of a larger sensory motor organ, the human "meta-genome", picking up and reacting to cues from its environment much like our nervous or immune systems;
2. that the human *body* should be understood as a ecosystem with multiple ecological niches and habitats in which a variety of cellular species collaborate and compete; and
3. that human *beings* should be understood as "super organisms" that incorporate multiple symbiotic cell species into a single individual with very blurry boundaries, like a colony of blue-green algae on a massive scale of complexity.[63]

Brigitte Nerlich and Iina Hellsten found that trends in public perceptions of microbes have changed over the course of the Human Genome Project (HGP) and the HMP. They found that a new language is emerging around microbes and their relation to humans. All told, they found that microbes are not seen solely as enemies, but rather as buddies or helpers, especially with the increase in probiotic use. They are a boon to human health. Crucially, "humans are no longer conceptualized as the pinnacle of evolution, standing apart from microbes seen as populating the bottom of the evolutionary tree: we are all human–microbe hybrids" and, in fact, "a new understanding of life itself seems to be taking shape: life has to be seen in a much less deterministic and much more fluid and flexible way."[64]

Metaphorical framings built from these understandings of microbial communities support system-based understandings of complex processes to body politics and politics more broadly. Conceptually, metagenomics implies that the human body is a communal gene pool, and its "self-extending symbioses," are "highly adaptive and robust against environmental perturbation" and "dynamic, self-sustaining and self-repairing processes."[65] Many problems that the state has been unable to address—warming oceans, pandemics, climate change, flows of immigrants and migrants—become easier to address when the emphasis is shifted to how complex heterogeneities respond to change. Myra Hird identifies this way of thinking as a "microontology" that understands humans to be "enmeshed in a complex web of co-domestications . . . such that natural, social and cultural selection may not be so definitively distinguished."[66] The micro materializes when metaphors summon it to the front.

In addition, this shift may suggest wider change in the way the human body is perceived. There is an emerging consensus that "concepts of the human body, the self and of what it means to be human are changing," but the extent to which this will affect human identity is yet unknown.[67] It may be that "established metaphors are also open for re-opening and re-interpretation, if, for instance, new data put the content of the metaphor up for debate."[68] Metaphor can *create* a reality rather than just explaining the current state. In other words, the ontological and corporeally based role of metaphor allows the awareness of the metaphor to lead to social and political change. Metaphors that are no longer fixed can be "building blocks used for making new discoveries."[69] It is crucial to recognize these modes of "concept visualizing" and that this provides the first step in transforming them" (67).

Insights from the life sciences and metagenomics can help us see other actants as vital to politics as well as our survival. It also aids in acknowledging, from both inside and outside our bodies, the potential inadequacies of the humanist idea of the individual. As our understanding of the complexity, porosity, and incorporation of the human body with other beings has grown, so, too, must our idea of the body politic. A "new anatomy of collective bodies" and the political ethics appropriate to it must be debated in as many forums as possible.[70] This interest in enriching, enlarging, and recognizing other actants in our global demos is coupled with the questioning of classical interpretations of sovereignty, law, and democracy. They may "continue to organize current political discourse—but their effective meaning always appears weaker and lacking any real interpretive capacity."[71] In Latour's words, "From now on, when we speak of actor we should always add the large network of attachments making it act. As to emancipation, it does not mean 'freed from bonds' but *well*-attached."[72] The loss in classical interpretations need not be cause for alarm, but rather taken as an opportunity to explore new forms of political community and belonging.

The language and affect involved in explaining the complexity of metagenomic discoveries to the public and the scientific communities help with these reconfigurations in the social fabric. This opening for creative revision brought by metaphors can shape politics with the help of additional actants playing more lively parts. By transgressing the boundaries of our autonomous selves and acknowledging the symbiont colonies in our bodies we can begin to see the ethical and social implications made possible by relations with diverse objects. Put differently, HMP and metagenomics can provide new concepts that are better equipped to supply IR with different forms of practice for sustainable, ethical, global living with one another, and with other life forms, as a *bodies politic*. These practices can operate at the micro- and macro levels through how IR understands the body and how these bodies exist together in political communities.

For example, the international system of sovereign states can also be understood as a site of cultural production, interactions, abstractions, and boundary creation. What is the state but an embodiment of ideas and beliefs about the space and place where politics is enacted and realized? To add the micro understanding of metagenomics and microbial communities allows for the human's commensal, host–guest relations to be a used as a dynamic model for human communities at the macro level. This model may demand that humans react differently to issues and scrutinize the effectiveness of current international institutions. Our responses to climate change, for example, have to take into account that our actions affect more than humans, and, importantly, that we function in a larger community with actants that support us and offer their own activities to the preservation of the earth's biosphere. In the words of the Committee on Metagenomics:

> Humans might survive in a world lacking other macroscopic life forms, but without microbes all higher plants and animals, including humans, would die. Not only can many individual

systems—for example, the human gut or such processes as the bioremediation of toxic hydrocarbons—be seen to be the tasks of complex and dynamic microbial communities, but these communities are themselves constituents of even larger systems, predominantly microbial, that collectively make up the biggest and most complex functioning system we know: the biosphere. Whatever the causes, extent, and consequences of the global climate change now upon us, the biosphere's response to the changes—and human survival—will depend on its microbes and their activities.[73]

Another way to imagine these lively bodies, or superorganisms, at the global level is through the idea of the assemblage. An assemblage is a collective entity made up of a gathering of things or people, but importantly, assemblages are created through relationality and never, writes Jussi Parikka, "from a prescribed relation hiding inside it, as if it were a seed."[74] There is no script, only self-creation through assembling and living together.

In genomic terms, a gene is not a book or a code that was hidden, and that we can now read through its sequencing, but, as shown earlier, a process that includes protein expression, environmental factors, and microbial relationships as well as vertically inherited traits. It is a "folding of the inside and the outside"; this folding is a material connection so it is "not a question of the body representing drives, forces, or even ideologies but of an intermingling of the world."[75] The organism is understood not as a reaction to a pregiven set of environmental stimuli, but as a response to the environment in which it is embedded and with which it must "harmonize" to survive.[76] Our environments guide us in becoming who we are, in a most material sense.

Symbiopoiesis

To better understand the global implications, a beneficial exercise lies in answering the question posed at the beginning of the chapter, "What would a microbe do?" with symbiogenesis and the refigurations of the human body that this chapter has sketched. I offer this provocative start to a new conversation about how new materialism and metaphor—alongside the recognition of microbial communities in the human body—can change perspectives. First, the human should be understood and celebrated as a superorganism who will interact with former strangers and new allies frequently to form stable, unique relationships.[77] Second, at the personal and institutional level, we will adapt with fluidity and rapidity to changing conditions while keeping multiple perspectives in mind (90). Third, it is vital to be a flexible, cosmopolitan entity, and see others as an ongoing and essential part of your transformation (90). There is no place for pernicious isolationism. Fourth, it is imperative to "form cooperative associations in a plurality of forms" with others. This creates conditions that allow us to be disposed to act for the benefit of others knowing full well that our survival is intertwined with others, and their well-being and thriving is our own.[78]

These answers can lead to two important implications for IR: a rethinking of global ethics and responsibility leading to a "diffuse sort of ontological gratitude for the post-human era, towards the multitude of nonhuman agents" who support us.[79] This diffusing, or flattening, of social action and ties into a continuum of dynamic object interactions between humans and nonhumans, states, bacteria and biomes, make the nested and imbricated nature of politics inside and between body politics more visible.

In this, ethics is understood as "what is to be done" to find affirmative practices and strategies for responding to a complex and messy world, rife with unforeseen challenges and actors.[80] A distinctive set of ethical and normative considerations travels to the forefront. Questions such as, "What does it mean to be human?" for philosophy, "How do we secure such a human?" for security studies, and "How do we respond?" for policymakers will take on new meanings and have new reference points.[81]

Ethics from this perspective is both more modest and yet more vital. Habits—from what we eat to what medical choices we make, from antibacterial soap to organ sourcing—can alter global networks and either improve or inflict harm on communities that heretofore were unseen. Communicative, anthropocentric, and rights-based ethics can only guide and inform the discussion so far in understanding the challenges and opportunities in the twenty-first century. Bioethics, with its focus on after-the-fact normative frameworks and protective regulation also falls short. The "ethical" needs to be a call to reimagine our relationships and receptively approach the world in all its dynamic complexity. Most important, to respond (not just react) with care to other forms of life, not as an Other or an outsider, but with a sense of responsibility to others in the world of which we are a part even if we may never meet them or know precisely what their claims of rights and wrongs may be. "If more people marked this fact more of the time, if we were more attentive to the indispensable foreignness that we are, would we continue to produce and consume in the same violently reckless ways?"[82]

Indeed, many pressing concerns exceed our political system and categories. Efforts should be made to scrutinize each and admit the possibility of other forms of life as we are *responsible for* and need to find a way to *respond to* a multitude of actants in a global demos. If, through research such as the above, we can no longer uphold the fiction of autonomous selfhood, then we must begin to think about the implications for institutions we create in our own image. These ethical, political, and metaphorical shifts will cut directly to the center of the notion of the liberal atomistic individual; the narrowly defined and rational *homo economicus* is challenged by the refigured and hybrid subject—the *homo contaminatus*—whose judgments may be altered by a wealth of unseen bacteria and viruses, commensal processes, and coevolutionary relationships. This hybridity can be used positively to understand how humans interact with, and alter, their lived environments as part of nature rather than divorced from it. In other words, theory and concepts must always be attached to the real in some way. A new grand, high intellectual theory will not help us. Instead we need those thoughts and concepts that are fleshy, dirty, attached, and molecular rather than relating to the whole. This allows for mingling that uses purity as a way to measure hybridity and borders as markers of transgression.

To take metagenomics and the microbiomes of the human body seriously means the human body becomes a community, not only a container—a "mutualistic human-microbial" series of interactions.[83] "We humans, like all organisms, live embedded in ecological communities ... each person, each dog, each tree is composed of many different living parts that can be detected and identified."[84] The HMP has helped to do this detection and identification.[85] From this perspective it is impossible not to see the similarities between relationships in the internal relations between members of microbiotic communities in the human gut and the relations between members of a political society. This will be addressed in the next chapter as the *lively vessel* is used as a base for reengaging with the metaphor of the state-as-person metaphorically framed as the *contaminated state*.

3

States in Nature, Nature in States

But now I sing not war,
Nor the measur'd march of soldiers, nor the tents of camps,
Nor the regiments hastily coming up deploying in line of battle;
No more the sad, unnatural shows of war.

—Walt Whitman, *Leaves of Grass*

Just as metagenomics and the Human Microbiome Project were used to complicate the body in the body politic, this chapter will use the processes of the immune system to address the state metaphorically as contaminated and diverse, rather than pure and homogeneous. The "personhood" of the state is made more complex by reciprocal relationships between nested sets of permeable bodies.

To begin to undo the metaphorical and ontological knot between war and unity—based on homogeneous, or "pure," national communities—I "contaminate" the state as person with difference and plurality through the idea of the parasite. The parasite does not always have a negative effect on health, but rather can invigorate the host's immune system. This claim, based on emerging research in immunological response to coevolved parasitic partners, is applied to the state imagined as a person. In this model, health must be defined as an ecological balance of diverse collectivities rather than the pathogen-centered elimination model. In other words, human health is not necessarily improved by killing that which is foreign, but by learning to adapt to, ignore, or live with that difference. If the health of the state as person, according to Lakoff's views, is through wealth, military strength, and industrialization as development—with undeveloped states as backward—health for the contaminated state will be based on different criteria. How this state-as-person enjoys and keeps good health is much different. The previous chapter asked, "What would a microbe do?" and answered that organisms give way to heterogeneous, flexible virtuosos then this chapter will query: "What is good health for the state?"

Biomedicine and evolutionary medicine supply the old friends hypothesis or the depleted biome theory,[1] and these will aid in defining healthy states. Proponents of the old friends hypothesis study the relationship between helminths—eukaryotic worms—and mammalian immune systems, and use this hypothesis to address a myriad of health concerns. Many of these health concerns seem to stem from industrial societies and the loss of long-term coevolutionary relationships with certain bacteria and parasites. They argue that modern medical practices and sanitation have removed certain microbial partners from the environment, which comprises the human immune system.[2] The removal of these commensal, quasi-commensal, and parasitic worms have adversely affected human health in some conditions.

Specifically, the activity of diverse microorganisms and the interactions between the host and microbes taught the human immume system "not to react to everything it came across."[3] In other words, the immune system needs something to react to or it will begin to "attack" itself. These medical studies suggest that a lack of microorganisms can lead to allergies, and it has been extended to explain disorders such as inflammatory bowel disease, autism, neuroinflammatory disorders, atherosclerosis, some forms of depression, and some cancers.

This is not to discount the creation of penicillin and vaccines,[4] or to argue that it is not useful in certain circumstances to use one metaphor over another in every case, but rather to look at how the metaphors used in the field of medical immunity were born from an idea of a host at battle with outside invaders. Later researchers would find that these metaphors were insufficiently based on partial knowledge and do not tell the complete story of a well-functioning immune system. A healthy immune system does not just keep foreign material out, it regulates a complex system of flow and exchange with the wider environment. Therefore, "a new paradigm is needed that incorporates a more realistic and detailed picture of the dynamic interactions among and between host organisms and their diverse populations of microbes, only a fraction of which act as pathogens."[5] With a broader vision of the innate and active immune system in mind, war metaphors are found to be insufficient and potentially harmful to humans and their environment when used to explain and respond to infectious diseases and microbial communities. New metaphors need to be "crafted" in order to explore the host–microbe relationship to better inform policies and further research.[6]

Symbiogenesis, introduced in the preceding chapter, works alongside the old friends hypothesis to destabilize the Darwinian focus on the struggle between organisms to a focus on long-term relationships between species. This, according to David Napier, can aid in envisioning a creative identity based on complexity and cooperation, rather than the transcendent, unilinear arguments based on natural selection.[7] The human immune system and the processes of immunity provide insights into political relationships writ large and into a further understanding of the body as embedded in an environment rife with long-term connections and transformations to and with other bodies, which further adds to reintroducing collective values through complex metaphors of cooperation.

State as Person

In 2004, the *Review of International Studies* journal hosted a forum for the debate around the "personhood" of the state.[8] It centered on Alexander Wendt's claim that it should be decided, once and for all, whether the state was *actually* a person or whether we merely spoke of it *as if* it were a person. In Wendt's opinion, the crux of the debate lies in deciding this issue. This is part of a larger conversation, as noted by Patrick Jackson in the forum's introduction: "A major debate in social theory for many years has involved the question of agency, and whether agency can be meaningfully located anywhere other than in constitutively independent human individuals."[9] Therefore, to locate agency in the state within this framework, the state must be a person to have efficacy, and even more important, subjectivity in International Relations.

The adoption of materialism and science studies—specifically Latour's "actant"—makes the above debate a moot metaphysical point. In Latour's flattened topography of social construction there is no need to differentiate between "agents" and "structures." If the state is taken as an actant then there is little need to decide whether it is an actual "person" or not since its ability to be an agent is not under scrutiny, but rather its efficacy as an actant that is being analyzed. The expanded form of Latourian agency allows for the state to be treated "as if" it is a person while still being able to be an actant. The "as if," or the metaphor deployed, directs the ways in which the state can act and respond to the "real" world. Thus, following Latour's Actor-Network-Theory (ANT), the question to ask about any agent is: "Does it make a difference in the course of some other agent's action or not?"[10]

In other words, the state's metaphysical status is not the issue, the question concerns the way in which we speak about its composition, or its ontology. The state's personhood necessarily disappears. There have always been complex bodies, rather than subject/objects or actor/structures. Therefore, the dualist language of agent/structure is not capable of making the construction of the international any clearer. We cannot fall back on it. What is needed is a durable vocabulary for a world in which the real is not entirely socially constructed or a priori. In other words, the choice is not between a vulgar realism, in which all is real, or a vulgar postmodernism, in which all is constructed, but rather lies in following the complex configurations that objects fall into during the processes of interaction and becoming. As with the biological body in previous chapters, the state's "being" is less important than its relationships with others in its environment. In a kind of radical constructivism, the state is created through the interactions it encompasses. This book presses IR to take its analogy of the "state as a person" to be taken seriously, perhaps even more seriously than IR has been asked to thus far, to find new possibilities for building global political communities.

If the debate thus far has been, "Is the state a subject like a person?" then, by diffusing or flattening social action and ties into a continuum of dynamic object interactions or translations, attention can be shifted to the durability and extension of these interactions, or "What does the state do?" The state represents translations and interactions between human and nonhuman bodies, courthouses, roads, laws, voting boxes, members of Congress or Parliament, licenses, treaties, and all the many things that make up the state as we know it in international relations. The state is not reified in order to explain other processes at work, nor is it just "social" or "ideational" as the constructivists would make it. Then the work becomes a matter of being able to understand how these translations fit into the wider process of community creation.

Of importance are when the state is spoken of as person and the interactions that follow from this. The metaphor itself is an actant in this situation and is interrogated as such. In fact, the focus on the metaphor is the "trick" to getting these associations to show up. "This is why specific tricks have to be invented to make them talk," Latour says, "that is, to offer descriptions of themselves, to produce scripts of what they are making others—humans or non-humans—do."[11] The way to begin an analysis of the state would not be to start from the state's metaphysical status, but to follow the actors that make up this network of translations, or the processes of making connections and making sense of the connections between actants. It is through translations that sociality is established and stabilized not in the hierarchical ordering of the different elements of a system.[12] This is a recording of the state that must stay dynamic like a live network connection rather than a prerecorded studio performance.

This chapter is interested in the production of the state as a territorially bounded entity based on a regime of inside/outside sovereign politics, and how this form came to be seen as natural and necessary; identity is constructed through these exclusions and seeks to make these constructions less facile by showing that the state, just like the human body, may not be what we think it is. Or, put slightly differently, the metaphors we have for speaking of the state as a rational, unitary actor may be hiding the actual processes that keep the state functioning. Our ideas of the state—and the body politic—are limited because of the way we understand bodies as autonomous sovereign actors in rationalist theories, ergo our politics are limited.

This limit is seen in the language of sovereignty and sovereign power, and although it has always desired to make life livable, this is imagined as setting boundaries that create ordered lines in the midst of chaos. The sovereign state, and later the nation-state, is a territorial instantiation of the belief that security lies in eliminating conflict inside borders with homogeneity, and the simultaneous creation of a dangerous and threatening outside. This ties the sovereign, the subject, the state, the citizen, the nation, and the human in a "tight existential bond" making security an ontological form, as Anthony Burke further explains, "based on the abandonment and repression of the Other."[13] In other words, the state operates on the logic of either/or and friend/enemy distinctions as a way to build community.

Communities—in whatever form they may take inside the state—are then identified as having a sameness within that is formed collectively through a similarity of "habitus" contrasted to a difference without that looms as a danger to this perceived unity.[14] By dissecting and isolating the world into discrete sovereign states, we are unable to see other collaborations that may be possible. This leaves us less able to care for the bodies, the rights, and the freedoms of those inside, and in-between, these states.

> If sovereignty is in some trouble, not least because political life does not always seem to be where it is supposed to be within the bounded authorities of a modern system of states, then this

trouble will be expressed in relation to the practices through which modern forms of sovereignty were constituted as attempts to respond to pervasive difficulties in defining what the world is, or must be, how this world must be carved up, how the carve-up might become authoritative, and what must happen at the boundaries (borders and limits, spatial and temporal).[15]

Our political categories are not matching the real and the limitations that follow from this mismatch need to be examined critically.

Aseptic Politics

With these preliminary points in mind, it is not difficult to metaphorically frame the state, as it is currently defined, as an example of the "pure culture" paradigm in biology. Pure culture is defined as a lab culture that contains organisms of only one type in a sterile liquid medium. "In the pure-culture paradigm, the presence of multiple species in the same culture medium means 'contamination.'"[16] From a biomedical and metagenomic viewpoint this paradigm has limited both what microbiologists have studied and how they think about microbes. The Committee on Metagenomics stresses that the pure culture paradigm is unable to see the dynamics of bacterial communities as it leads studies to focus on what a microbe *is* rather than what it *does* within a community; therefore, they are ultimately unable to fully understand the microbes they are studying. In addition, "microbiologists learn early in their training the tricky job of keeping all other life out of a pure culture by using aseptic technique." "Aseptic" can be translated as "without contamination" and it indicates that this technique "requires that a microbiologist manipulate cultures without letting in any unwanted bacteria."[17]

Just as the human body is a nested set of permeable bodies, the state, too, is seen as a body composed of multiple relations. State politics based on the inside/outside dichotomy are critiqued as a kind of "pure culture," and explored metaphorically by using the notion of the "depleted biome." By using the term *inside/outside*, I draw on the principle of state sovereignty as an account of the nature and location of political community. "Specifically," R. B. J. Walker writes, "the principle of state sovereignty offers both a spatial and temporal resolution to questions about what political community can be.... It is a very powerful, even elegant answer to the deeply provocative question as to how political life is possible at all."[18] Importantly for this book, the territorial state allows for a clear distinction between those that are inside the borders of the state and those that are outside the borders and therefore excluded from community, ethics, and the ability to pursue the "good life" within the state. These clear territorial boundaries often lead to an "ethics of absolute exclusion."[19] These exclusionary politics—including metaphorical entreaties to purity inside the state—or the "pure culture" model—often lead to extreme forms of violence such as ethnic cleansing, genocide, or "lethal projects of social engineering intent upon eliminating certain undesirable and 'contaminating' elements of the population."[20] Metaphors based on the perceived social health of the state through purity are often used to justify authoritarian and fascist political programs.[21]

To return to the creation of security within the state, the resonances with "pure culture" and "aseptic techniques" become clearer. Unwanted bacteria are eliminated to protect purity and within the state anything that is seen as a threat to the state's integrity—minorities, critics, reformers, immigrants—are eliminated or expelled.[22] The "pure culture" paradigm has an intimate relationship with a pathogen-centered focus in immunology. "Causes" of contamination are identified and eliminated through aseptic politics like anti-immigration and sedition laws, constriction of free speech and assembly, and radical violence such as lynching and ethnic cleansing. Other social effects of this pathogen-centered, aseptic politics can be identified. It aids reinforcing lines between "us" and "them" on micro- and macro levels, it can excuse and permit violent reactions to anything that is different or marginal, and this violence and attitudes of victimization "[affect] what we believe to be socially and scientifically possible, and even more important what we believe to be socially acceptable."[23] Biological ideas, combined with the modern state's understanding of modernity as a cleansing project, were based on purity and used to justify racial oppression and discrimination, and in World War II these ideas were used against the handicapped and homosexuals.[24] The twentieth century saw the death and immiseration of millions of people.

> With the rise of the nation-state and its imperialist and modernizing ambitions, tens of millions of "backward" or "savage" indigenous peoples perished from disease, starvation, slave labor,

and outright murder. Sixty million others were also annihilated in the twentieth century.... The list of victim groups during this "Century of Genocide" is long. Some are well known to the public—Jews, Cambodians, Bosnians, and Rwandan Tutsis. Others have been annihilated in greater obscurity—Hereros, Armenians, Ukrainian peasants, Gypsies, Bengalis, Burundi, Hutus, the Aché of Paraguay, Guatemalan Mayans, and the Ogoni of Nigeria.[25]

Likening the state to a "depleted biome" using the logics of pure culture bares the violence inherent in these ethics of political community and helps to rebuild a link between plurality and difference as a necessary ingredient in healthy, diverse communities, just as diverse microbiotic communities keep the human body functioning.

Immunological Frames

Theories of immunology represent ways in which we perceive the human body interacting with its environment and how organisms coexist with each other. "The development of immunology represents the convergence of several fields: microbiology, pathology, embryology, evolutionary biology, and biochemistry."[26] In addition, immunology and the history of immunology have become a contentious field in part due to the pressing concerns to which it is responding. What was once a relative backwater in the field of medicine became an important area of research when HIV/AIDS entered the global scene.

In medical terms, immunity is the process of preserving the life of a body. The scientific community borrowed the term from the legal concept of *immunitas:* a Roman legal term describing the status of a person or a body as free of legal obligations such as liability for damages or prosecution of criminal acts. [27] In fact, biological immunity as it is known today was born of the late nineteenth century with Elie Metchnikoff and his phagocytosis theory of host defense. The dominant perception of these processes is one of defense and intrusion of a host by invading pathogens, as it focuses on the processes of phagocytosis, or when living cells engulf other cells; it is chiefly a defensive reaction against infection in animals. In part, the focus on this portion of the immune system was born due to Darwin's natural selection theory. Tauber and Chernyak attribute this to the way in which Darwin's thinking profoundly changed the concept of the organism. The organism lost its meaning as a *balance* of forces or humors and became the product of *imbalanced* structures and functions that adapted in a competitive, hostile environment: "Explicit appreciation of this basic shift in the metaphysical structure of the organism was made by Metchnikoff, who presented the phagocytosis theory—the basic conceptual notion of immunity—in response to how the organism was defined by such evolutionary challenges; in the process, he established a modern definition of selfhood."[28]

Competition was the guiding understanding: "Metchnikoff's vision of the phagocyte was the epiphany of the striving organism seeking its advantage in competition with its environment to establish hegemony in one sense, harmony in another."[29] This has led to a continued preoccupation with warlike interactions and has emphasized the role of "discomfiting military language that highlighted tracking of enemies, camouflage, battles within, attack rates and victorious outcomes for one or the other combatants."[30] As discussed previously in regard to interventionist language and medicine, the war metaphor in immunology continues to inform and feed medical treatments and military action; both cultures learn and apply these images metaphorically and symbolically. Donna Haraway writes:

> For example, arguing for an elite special force within the parameters of "low intensity conflict" doctrine, a U.S. army officer wrote: "The most appropriate example to describe how this system would work is the most complex biological model we know—the body's immune system. Within the body there exists a remarkably complex corps of internal bodyguards. In absolute numbers they are small—only about one percent of the body's cells. Yet they consist of reconnaissance specialists, killers, reconstitution specialists, and communicators that can seek out invaders, sound the alarm, reproduce rapidly, and swarm to the attack to repel the enemy.[31]

Although it was recognized even at Metchnikoff's time that some bacteria did not cause disease, it was not fully appreciated until the advent of immunosuppressive therapies in the twentieth century. But by this time, the idea that medical research was based on the search for the cause of microbial disease and antibiotic ways to "eradicate" these causes was firmly in place, and a more nuanced, ecological approach was a relative backwater in immunology.[32]

The medical effects of this framing has led to antibiotic-resistant bacteria like Methicillin-resistant Staphylococcus aureus (MRSA), New Delhi-metallo-beta-lactamase-1 (NDM-1), a bacterial "superbug,"[33] and a novel gonorrhea, H041 found in East Asia.[34] Different microbial threats have also emerged and reemerged since the optimistic belief that humanity had won a war against pathogenic microbes prevalent up until the mid-1960s.[35] Legionnaire's disease, toxic shock syndrome, AIDS, Lyme disease, West Nile encephalitis, H1N1, reemergent influenza, Ebola, hantavirus, and SARS are among them.[36]

All told, these medical effects led the Forum on Microbial Threats at the Institute of Medicine (IOM) to collect a variety of individually authored papers and commentary, with the goal of replacing the war metaphor in infectious disease intervention. This report represents an expanded approach to microbial threats to health that emphasizes a change in perspective in order to recognize "the breadth and diversity of host-microbe relationships beyond those relative few that result in overt disease."[37]

Advances in computer technology—as with metagenomics and its sequencing machines—and improved statistical methodologies aided researchers in gathering new information about the wider role of the immune system. These technologies revealed the "intricate biological networks" and allowed for "sophisticated diagnosing of medical maladies." These technologies also aid in showing the immune system as consisting of molecules and cells connected in a network and place special emphasis on "the interchange between the organism and its environment, the processing of information, and the regulation arising from responses to this larger context."[38] It can now be seen that the immune system functions at the interface, or junctural zone, between an individual organism and its environment, but that these operations can be understood as a *dialogue* between the two in "response to the challenges received from diverse encounters" rather than a violent altercation. In short, the immune system functions at the "interface of host organism and its environment both defensively and cooperatively."[39] It is a two-step process in which the immune system cognitively perceives its environment, and then responds in a manner deemed adequate to its perception.

By framing the immune system as being an actant in this environmental exchange, "regulation becomes a process arising from both internal equilibrium mechanics *and* stimulation from external sources" (232). It can then be regarded as a first line of defense, and more broadly as an "information processor" for the host. To understand this process through open cognitive functions allows for open information flow and gives an infrastructure that can support an ecological orientation. Likewise, "the wider reference of 'ecological immunology'" can better understand how "a particular antigen might be regarded as harmful to a particular individual or species (and thus subject to immune destruction)," and also "determine the costs of defensive mechanisms to the community-at-large" (232).

More generally, new understandings of immunology also show an interest in modeling the immune system as an open, or holistic, network rather than a closed one, and join a similar interest in many fields to use the insights of systems analysis to adequately address complex, dynamic systems. This allows new metaphors, or organizing orientations, from ecology, cognitive processes, and systems biology to begin to take shape.

An intervention into the language, and the connections between the broader discourses of biology and the social sciences, is productive. It does not take the form of a criticism of science or the relationship between the discourses, but rather focuses on the opportunities for this borrowing of analogies and metaphors to affect our social and political definitions. The language that we use and the models we create, as scientists or academics, can provide clues about and explain different aspects of political organization. These new models of the immune system based on open networks and cooperation bring different aspects of "interdependence, co-existence, and mutual determination between discrete living beings" to the forefront. These, as Laurence Whitehead says, "may all have something to tell us about how human political communities can persist, adapt, and even govern themselves without self-destructing."[40]

The forum, in the report discussed above, says that in response to these changes in understanding the immune system, scientists should recognize that a "new paradigm is needed that incorporates a more realistic and detailed picture of the dynamic interactions among and between host organisms and their diverse populations of microbes."[41] This paradigm will place disease in an ecological context that "promotes a more realistic, deeper, and nuanced understanding of the relationships upon which these systems depend" (2), and this, in turn, aids in focusing on the interactions of various elements of the immune system within its milieu.[42] Further, this ecological view "recognizes the interdependence of hosts with their microbial flora and fauna and the importance of each for the other's survival."[43] This recognition demonstrates the medical community's interest in using and understanding the potential of nonpathogenic communities, both to improve health and to understand the change from commensal to pathogenic. Likewise, this broader view of immunity must take into account how pathogens coexist within host microbial communities. The human gut, in this book and in the forum's report, is exemplified as a community in which very few relationships are pathogenic.

The forum also asks the medical community to admit a vulnerability to both emerging diseases and those we thought had already been conquered. In the face of the challenges created by emerging diseases—both viral and bacterial—and antibiotic resistant bacteria, the forum writes that the war metaphor is characterized by the systematic search for the microbial "cause" of each disease, followed by the development of antimicrobial therapies. The war metaphor's pervasiveness diverted medical and scientific research from the "adaptive role that microbial agents in the functioning of the host organism" and "in a very real sense the triumph of the war-metaphor sidetracked the evolutionary and ecological role of microbes, and seduced the medical community to believe that health could be gained through winning a war."[44] This belief that we were "winning" was further enforced by improvements in modern sanitation, diet, and living conditions in many areas of the world.[45]

This paradigm can no longer guide biomedical science or clinical medicine because it is insufficient to answer these challenges. The language of war is insufficient to fully describe the well-functioning immune system, and the consequences are too dangerous. In other words, if humans continue to see themselves at "war" with these bacteria, humans will ultimately "lose" the war—even if battles have been won in the past. The forum begins the report with this quote from the World Health Organization:

> 2000 B.C.—*Here, eat this root.*
> 1000 A.D.—*That root is heathen. Here, say this prayer.*
> 1850 A.D.—*That prayer is superstition. Here, drink this potion.*
> 1920 A.D.—*That potion is snake oil. Here, swallow this pill.*
> 1945 A.D.—*That pill is ineffective. Here, take this penicillin.*
> 1955 A.D.—*Oops... bugs mutated. Here, take this tetracycline.*

> *1960–1999 — 39 more "oops." Here, take this more powerful antibiotic.*
> *2000 A.D. — The bugs have won! Here, eat this root.*[46]

As the glib quote above demonstrates, focusing only on defense can, paradoxically, make us unhealthier and be far more dangerous than it seems at the outset. The immune system is not at war with elements outside the body. Defense is part of the story, but defining the organism and its relationships with its environment is equally important. "Whether cooperation or war should ensue would depend upon whether or not the conflict-mediating mechanisms functioned properly."[47] The next section pairs this look at biology with the way in which IR understands the state as a person and the debates around its agency in the international.

Exposing the Body

The old friends hypothesis, the microbial exposure hypothesis, or the biome depletion theory (formerly, the cleanliness hypothesis) are all terms created to theorize the rise in allergies and autoimmune disorders in industrialized countries since the 1980s. The name change occurred because of confusion over the word *cleanliness,* which misleads one into thinking that hand-washing or overall cleanliness plays the most important part. This suggests to the general public "that we should abandon hygiene, and that what we have developed over centuries — all the efforts of all medical researchers and doctors — is wrong and should be dropped." The new terms "return the focus to the role of the microorganism in educating the immune system."[48]

Also, these theories do not argue that biome depletion or lack of helminths in the immune system are the only explanation for immune disorders, but that they have "considerable explanatory power."[49] The increase can also be "connected with genetic and epigenetic factors that predispose to disease as well as environmental 'triggers' which initiate pathogenesis."[50] To add to this, "the identification of genetic factors or a 'trigger' (often a viral infection or some other environmental stimulus) does not invalidate the applicability of the Biome Depletion Theory to a particular disease. Rather, it is the incidence of a particular disease in post-industrial countries and in developing countries, in conjunction with the presence of an immune component of disease, which is indicative of the involvement of biome depletion in pathogenesis."⁵¹ Epidemiology supports the conclusions of this theory, but there is still disagreement on the mechanism that causes infections to regulate the immune system, so the hypothesis remains controversial.[52]

Simply defined, this theory says that immune system development is limited by less exposure to infectious agents or microbial communities that, in turn, increase susceptibility in humans to allergies and autoimmune disorders. William Parker of Duke University Medical Center explains:

> This theory, in its present form, describes the medical impact of separating us from our partners in co-evolution by means of widely appreciated medical care in combination with technology that we now take for granted, including running water and toilets. The Biome Depletion Theory embodies the central idea that epidemics of immune-related diseases associated with post-industrial society are due to a pathologically over-reactive immune response, which in turn is caused by a loss of components of our biome that normally interact with our immune system.[53]

Increases in certain diseases are not seen in the developing world where more parasites are present, and therefore, there is greater exposure to microbial communities. These various "old friends," or the parasitic worms and microbes once prevalent in our bodies and environments, have almost disappeared because of changes in lifestyle. These include lactobacilli, saprophytic mycobacteria, and some parasitic worms, or helminths.[54] This is not to say that helminthic infections are not a common and serious problem, but the relations between humans and their guests vary widely. They range from benign to very serious and must be understood in context.

As one of the researchers asserts, "What we're talking about really is fundamental changes in lifestyle, it's not just the trivial matters of everyday domestic hygiene. It's the fact that we no longer drink water from the stream and we no longer have worms."[55] Medical researcher Joel Weinstock told Discovery News, "Many of these worms are bio-engineered for humans.... We adapt to them; they adapt to us. It becomes like an organ, just like your heart, your spleen or your liver."[56] For example, whipworms have been found to have a "profound symbiotic effect on developing and maintaining the immune system."[57]

A helminth, like the whipworm, is tolerated through immunoregulation.[58] Once established, these organisms must be tolerated (rather than registered as harmless). Through long-term residence in the human gut, the worm "taught" the "immune system not to react to everything that it came across."[59] As with the other members of the gut microbiome, like lactobacilli, the immune system came to "overlook" them over the course of evolution. Overlooking coevolved symbionts aids the immune system in not overreacting to foreign matter, or what it perceives to be foreign, in the case of autoimmune disorders. The focus on host defense has led to incorrect perceptions of how the immune system functions the majority of the time. "People get the immune system the wrong way around. We're so focused on the immune system responding to things, that we forget that 99.999% of the time, its job is to not respond to things. There's you and your breakfast and your gut, for a start. That's a lot of stuff to not respond to."[60]

Without these helminths, our immune system is increasingly "on the attack." Put differently, without these "others" in our system, our bodies increasingly become less tolerant and more prone to attack itself, leading to a decline in overall health. The body can no longer distinguish between what is a real threat and what is benign, such as between ragweed and food, or "perhaps because they [immune processes] are not busy fighting real threats, they overreact or even turn on the body's own tissues."[61] Of course, this theory does not support a return to premodern sanitation or desire elevated mortality or reduced life spans. Rather, the call is for a more nuanced approach to the issues.

One such nuance is the metaphorical conceit of *contaminated state*, an imago of the "person"-hood of the state.[62] This way of thinking can point toward new definitions of health for the state based on expanded notions of health through plurality. The state's health, as defended through ecological immunity theories like the old friends hypothesis, is different from that created through current rationalist theories; plurality and stimulation, not autonomy and purity, lead to strength.

This aseptic politics has left the state with health issues associated with the depleted biome theory. The state can no longer distinguish between what is a real threat and what is benign, as evidenced, for example, by domestic policies like the USA PATRIOT Act enacted in response to 9/11. It cannot understand that not all threat, come from the outside and, further, that not all difference found internally is dangerous to the health of the polity. Put another way, sameness does not exist within, and difference is found both internally and externally. The search for a singular, unified entity only does violence to this deep plurality. As is evidenced by the old friends hypothesis, neglecting and destroying difference, or relationships based on long-term connections, is ultimately unhealthy. Highlighting the contaminated and plural nature of the body makes the easy exclusions of the state, based on a mistaken belief that difference creates conflict and sameness leads to harmony, less facile. The Human Microbiome Project, described in the previous chapter, and ecological immunity theories show that a healthy human body is constituted through diverse relationships with multiple Others, both on the inside and the outside, and this can be used as a model to create new strategies that secure collective identities.

The body politic, as instantiated in state practice, is made more complex by the creative revision allowed for in ecological immunity theories, which are based on networks and cognitive processes. Specifically, ecological immunity theories find the immune system to be more reciprocal and open than combative and closed. Haraway writes:

> The genetics of the immune system cells, with their high rates of somatic mutation and gene product splicings and rearrangings to make finished surface receptors and antibodies, makes a mockery of the notion of a constant genome even within "one" body. The hierarchical body of old has given way to a network-body of amazing complexity and specificity. The immune system is everywhere and nowhere. Its specificities are indefinite if not infinite, and they arise randomly; yet these extraordinary variations are the critical means of maintaining bodily coherence.[63]

The immune system is formed by the complex processes of a body living in relations with communities, or by evolving "as an interface with symbiotic organisms more so than as a defense against invading organisms, although defense against invading organisms was almost certainly a part of that interface."[64]

Contaminated State

Just as parasites are often necessary for overall health in the human, the idea of the parasite as an "Other" is necessary for healthy politics. As discussed in the introduction, contamination is often needed to stay healthy and the above discussion on our coevolution with eukaryotic species has borne out this assertion. Caroline Hadley further explains that better health and lower risk of allergies can be found by understanding the role of "certain organisms that have, over the course of evolution, trained our immune system to be more tolerant."[65] It is now time to bring these elements together into a discussion of politics of contamination. To do this, I bring together both theoretical ideas of the parasite's role in production as a "noise" that disrupts the status quo, and the actual parasite as it is understood by the old friends hypothesis and ecological theories of immunology. The parasite brings a positive critique of the dualisms that structure the human body and state politics, and is offered as a way to articulate an alternative to political subjects created from these dichotomies based on politics of inside/outside.

It is important to note that the word *parasite* itself is a misrepresentation; parasitic relations differ from common usage and understandings of the word *parasite*. Parasitic relations can no longer be understood as a simple one-way relationship wherein only the parasite benefits in "coming to the table," with the relationship ending by the death of the host. This is due to the above findings of immunologists about the role of the parasite in a healthy immune reaction to outside stimulus. This revised understanding can be better defined by tracing the etymology of the word itself. It comes from the Greek, *parasitos,* meaning "beside the grain." This reflects that the parasite "was originally something positive, a fellow guest, someone sharing the food with you, there with you beside the grain."[66] In a simple definition by Michel Serres, "to parasite means to eat next to."[67] It is later that "parasite" came to have a negative meaning as someone who enjoys a seat at the table as a perpetual dinner guest, an "expert at cadging invitations without ever giving anything in return."[68] To call anyone a parasite is strong language marked by derogatory synonyms including cadger, bloodsucker, sponger, bottom-feeder, scrounger, freeloader, and mooch. In this definition there are no guests sharing food, there are only enemies and interlopers who are unwelcome and to be fought against.

This book cannot share this negative view of parasites; a more complex view must be taken to combat the affective power of the war metaphor. At the risk of overusing the dinner guest metaphor, the parasite repays the host for the meal with stimulating conversation and necessary news from the outside world. "The parasite is invited to the *table d'hote;* in return, he must regale the other diners with his stories and his mirth. To be exact, he exchanges good talk for good food; he buys his dinner, paying for it in words. It is the oldest profession in the world."[69] In terms of the body, the parasite stimulates the immune system thereby allowing it to better discriminate between harmful and benign. To "parasite" falls in a continuum of symbiotic relations between benign commensals, like lactobacillis, and the more harmful eukaryotes. As a patient undertaking helminthic therapy explains: "It's not healthy to lead an antiseptic lifestyle, and deadly parasitic diseases like malaria don't do us any good, but there is a middle ground, and an alliance to be made with microorganisms on the gray scale."[70]

Ecological immunity and the old friends hypothesis bring this more nuanced view of parasites to politics. Parasites bring much needed "conceptual twists" to discussions of community and collectivity.[71] I will highlight three ways in which the parasite adds this "twist." First, the parasite has a destabilizing effect on subject creation, and this changes the form of our collectivities. It helps move the discussion from the subject/object distinction, or the language of dualism, to the realization that relations are intersubjective. As explained by Joost Van Loon, "rethinking community through the figure of the parasite allows us to steer clear of both the survivalism of the solitary-autonomous but authentic individual and the mediocre identity politics of the herd collective" (252). To add to this, Michel Serres says that we must face this intersubjectivity head-on as a constant: "There is no system without parasites. This constant is a law."[72] Understanding this form of production—that it is "parasites all the way down"—is critical to finding the political in host/guest relationships. The strongest position and the weakest position are interchangeable and contingent in productive relations, or as Serres says: "The host/guest is universal, eater of all and eaten by all" (33). In other words, parasite politics can help to define our relations to production and consumption of resources in our communities as fluid and reciprocal. Every community is a community of parasitic symbionts where the contributions of members can change from moment to moment, day to day, year to year. This highlights the parasite as something other than an either/or choice of construction or destruction, order or disorder. Rather, it allows for fluctuation and contextuality.[73] We are parasites all the way down.

Second, the parasite can produce change in a homogeneous system by disruption, or as Serres terms it, "noise." This is an interesting play on the translation of the word *parasite* as noise or static. Serres writes, "*Static,* in English: parasite." [74] The parasite is an "operator that interrupts a system of exchange." It is not, as Serres argues, because the parasite takes without giving back to the host, but that the parasite stimulates the system as a whole. This noise, or static, disrupts the message and at first seems negative; the parasite can be viewed as a malfunction or disruption of a message. The parasite, in this view, would be an unwanted addition that should be expelled. This, at first pass, "elicits a strategy of exclusion" from the host but, as David Bell argues, this would ultimately be a mistaken response. In order for a system "to function to perfection," he writes, "it would seem necessary to eliminate all parasites," but this is not the case in complex systems. With the parasite as a constructive agent, and "by experiencing a perturbation and subsequently integrating it, a system can pass from a simple to a more complex stage."[75] In ecological theory, this is called the "intermediate-disturbance hypothesis." This theory holds that biotic diversity will be greatest in communities subjected to moderate levels of disturbance.[76] This can be viewed as an active process of self-organization that is autopoietic in character. Further, parasite, or "contagion" more broadly, "involves a micro-level transmission that activates a chain of individual responses, which reverberates through the body politic, destabilizing the pre-existing homeostasis."[77]

Third, the parasite introduces a different ethos to political relations. Van Loon stresses that "in becoming-host, the parasite is an effective ethical Other that can aid in engendering a sense of 'community-in-difference.'"[78] Specifically, parasitism engages ethics on a plane different from that of survival. It is not the survival of a particular symbiotic community that is at stake, but its transformation, its "becoming-other." "Hence on a very immediate and direct plane we can see how communities are formed on the basis of endemic parasitism," but to survive, the community must surrender to an Other, or the parasite (250).

Through the parasite and the old friends hypothesis, I argue that, contrary to much thinking in IR,[79] difference within cannot be negated and erased or serious consequences will follow. The human body depends on the bacteria in the intestine, and these commensals—including what we term "parasites"—depend on us. If these relationships ceased, both the human body and all its communities would die.[80]

Even more profoundly, it is this difference itself that creates healthier bodies. Many health problems have emerged because of a "modern obsession" with sterility equated with cleanliness and goodness. Although this was a consequence of the immediate benefits of modern sanitation, it now needs to be recognized as a "cultural artifact."[81] The cultural prejudices against worms and parasites should also be reassessed in light of the new information introduced by ecological immune theories. Parasites can no longer be seen as "despicable" and "vile," but understood as both parasitic and beneficial. Definitions need to shift to understand the complex forces at play that contribute to health. As Sharon Shattuck, director of the documentary *Parasites: A User's Guide,* asks: "If helminths were called probiotics instead of parasites or worms would people view them any differently? Who's afraid of the big bad worm, and at what cost does this fear come?"[82]

Metaphor has been used in this book to clarify what metaphors *do* rather than what they *are*. Therefore, an important question that follows is not simply, "Are there different metaphors that can be employed?" but rather, "Are there better metaphors?" I put forward the contaminated state as a metaphor that changes the focus of state policies as reflecting a need to sterilize and purify to the need to pluralize and understand productive relationships of give and take in a different way. The "person" of the state as contaminated can aid in defining the "person" of the state by its composition, and this composition as multiplicitous, diverse, contaminated, plural, and parasitic. The health of this state is markedly different from the current understanding through simplistic metaphors based on violent conflict and warlike interactions. More appropriate metaphors are needed to understand and change policies based on these understandings of health.

Immunology can take the discussion about borders and flow across borders in productive directions. To return to the parasite and the contaminated state, this perspective supports deep pluralism and a way to "engage with the other-within as part of a continuing process of transformation."[83] Although contra many of the tenets of Darwinism, this is also not a Lamarckian idea of evolution as inherited characteristics, but rather of "inherited bacterial symbionts" that merged because they infected one another and were "reinvigorated" by these incorporations. Thus, symbiosis has a "filthy lesson . . . to teach us: the human is an integrated colony of amoeboid beings, just as those amoeboid beings (protoctists) are integrated colonies of bacteria. Like it or not, our origins are in slime."[84]

If applied to the debates surrounding immigration, for instance, a different picture of the immigrant appears. Disease and foreigners have often been linked, and immigrants are often referred to as pests or parasites on the state. Calling immigrants "pests" and "viruses" makes it impossible to analyze the difference between foreign capital and the labor of immigrants, for example. It misunderstands that we need contamination and the "infection" of a plurality of outsiders to stay healthy. A petri dish with a sealed bacterial community within it will die from the center out; it needs an open border to thrive.

What metagenomics and immunology can offer this debate is the ability to accept that "contamination" is needed to keep healthy. Looking through these biological perspectives renders many processes transparent that we could not see before within the "depleted biome" of the modern state. We acknowledge flows in many other places and forums—such as intellectual knowledge, capital, scholars, professors, and doctors from other countries. Simply put, if we take the immigrants out of the state it is like removing the microbes from the gut. Just as we need different bacteria in our bodies, we need different people in the body politic. In these metaphors, the ontology supported is one that renders actants as worthy of care and regard by understanding the place of difference not as threatening, but necessary. Following these perspectives, the state would have to be adaptive and flexible in different ways to benefit from the flows and plurality within its borders. It would understand that governing as a contaminated state involves multiple stages of sensing, adjusting, and configuring reactions based on specific contexts (emphasis added):

> *The border,* or *ecotone,* is a transition site where co-mingling gives rise to unique dynamic relationships and innovation;[85]
>
> *Competition* will occur, but it is understood as a path toward unique opportunities that must be met in a shifting environment of friend and foe. This is not simply about identifying the Other as threatening, dangerous, and external and the state cannot define itself by "indolent innocence or persistent aggression."[86]
>
> *Symbiosis* can result in a complex new form, but must be understood in stages that cannot be described through warfare metaphors, but rather understands that information processing is not a matter of elimination and selection, but of convergence and transcoding.[87]

Healthy community creation "does not always mean removing the cause of the disease and expelling the carriers of germs,"[88] but rather in understanding the complex interplay between commensal and pathogenic relations.

Parasites are not always organized to the disadvantage of other citizens because "health has more to do with symbiosis than extermination . . . so disinfection and isolation alone will not contain the danger."[89]

This focus, or new metaphorical framings of health based on relationships, changes what can be pursued as national interest in the name of health. The state-as-person metaphor, as it is applied now, is used to justify all sorts of acts and policies when in fact "what is in the 'national interest' may or may not be in the interest of many ordinary citizens, groups, or institutions, who may become poorer as the GNP rises and weaker as the military gets stronger."[90] In addition, a vibrant democracy is the one best able to domesticate threats rather than survive behind an artificial "cordon sanitaire," and unhealthiness often arises from a breakdown of harmony or dysfunctional interactions between organs.[91]

The implications of these conceptions and metaphorical framings are far-reaching for both philosophy and biology, and will alter the tenets of both. The following chapter will gather these musings around immunity into a larger tale about the possibility of posthuman, complex, and life-oriented methods of study in the field of International Relations.

4

Posthuman Politics

I celebrate myself, and sing myself,
And what I assume you shall assume,
For every atom belonging to me as good belongs to you.

—Walt Whitman, *Leaves of Grass*

Whereas the first two chapters created a framework for thinking of familiar subjects in International Relations (IR) as strange, and the next two used that scaffold to support a pair of metaphorical conceits for theorizing politics at the microscopic, corporeal register to evince different affects, this chapter must necessarily widen to analyze the equally important macro view of the body as it is enmeshed in the biosphere and in preexisting definitions of life, writ large. In contrast to the rational, atomistic individual and the secure, aseptic state, the lively vessel and the contaminated state facilitate a sense of connection to the world through the processes of immune response and microbiotic communities. This, I suggest in chapters 2 and 3, nurtures a sense of connectedness within and among multiple bodies; relationships between and among nested sets of permeable bodies. New biological metaphors complicate the easy reliance on the body politic as a static entity that escapes evolutionary and historical change. To draw on another scientific field, a quantum physics of entanglement, of spooky action at a distance, rather than Newtonian understandings of classical mechanics, may explain social life with greater perspicuity.

Our atoms, as Whitman says above, belong to all. As Carl Sagan famously stated in an episode of Cosmos: "The cosmos is within us. We are made of star-stuff. We are a way for the cosmos to know itself." This is a most crucial change of emphasis for creating new designs for global thriving: to generate analogies for bodies to thrive in entangled communities and politics. This will mean finding new guiding fictions beyond the story of the social contract and the Westphalian state. The task becomes to expertly reverse-engineer from the small to large, bacterial to atmospherical, a refiguring of our political structures and institutions.

This includes a temporal element: to think from the future we desire to inhabit the actions we can enact in the present.

Shifting Metaphors

Historically, IR has been concerned with predicting futures. Many in IR long to bring it closer to the "hard" sciences and endeavor to create predictive theory based on positivist models of scientific testing and validity. The quantitative rules over the qualitative and often in reductive ways. It is said that IR suffers from "physics envy," and the "quest for predictive theory," as Steven Bernstein et al. argue, and that this most often rests "on a mistaken analogy between physical and social phenomena."[1] This model draws most of its strength from a Newtonian view of the universe as gross matter, with atoms as stable and predictable substances.

Newtonian physics conceived of a world of clocklike regularities that could be discovered through deductive theory and empirical research. Prediction was a reasonable goal because many of the phenomena studied by eighteenth- and nineteenth-century physicists were the result of a few easily measurable forces or of interactions among an extraordinary large number of units that gave rise to normal distributions.

After Einstein, and with continued advances in quantum physics, we are "no longer able to rely on the obsolete certainties of classical physics."[2] Particle physics changed scientific ideas about the composition of matter, and complexity theory continues to modify our understandings of the patterns and characteristics of matter's movement. All told, this has undermined our ideas—and models built on these ideas—of stable and predictable substances, and focuses on the world as complex, unstable, fragile, and interactive (13). Especially for International Relations, it is often impossible to assign metrics to variables that are deemed important and "as physicists readily admit, prediction in open systems, especially nonlinear ones, is difficult, and often impossible."[3]

Systems biology will be introduced as a possible partner to aid in understanding international relations as a complex system. This systemic understanding will be joined with a plural posthuman subject. The posthuman, as a concept, complements the book's approach to the human as a complex, organic assemblage of human DNA and multiple bacterial and symbiotic others. A natural conceptual ally exists in the idea of the posthuman.[4] We need new stories to tell its tale; biological and global social systems share conditions that could be better understood by methodologies that can move between scales and change based on dynamic conditions and emergence. As discussed in the introduction, bringing together the social and natural sciences through overlapping, or superimposing, the world of people and the world of things, adds a crucial element to a more malleable understanding of complexity in global politics.

Through metagenomics and immunology, biological metaphors provide IR with "more flexible and appropriate" analogical structures to highlight the "complex, open-ended, partially reversible, process of political construction" that exists globally.[5] Biological analogies allow for emergence, change, and autopoiesis in ways that are more sensitive to "soft" actual bodies as opposed to "hard" bodies of physics. Biology can also challenge IR's belief that prediction should be the only goal of any theory. As modern biology is concerned with evolution and cooperation over time between diverse communities, this viewpoint aids International Relations in thinking about how the past can apply and teach us about the future.

Specifically, by substituting biological metaphors for the current metaphors used in IR, the characterizations of global politics and the way these characterizations can make clearer the narratives in which our political concepts appear to be true. The relationship between medicine and intervention, and between immunity and community provides evidence that there is more than just a "reciprocal influence between the life sciences and the social sciences," and, continues Laurence Whitehead, that "multiple methods of investigation thus deserve encouragement . . . since all creative thinking involves the imaginative re-interrogation of established assumptions."[6] Whitehead identifies several ways that biological metaphors can offer aid in the social sciences:

> *Change.* They can identify regulatory principles that generate change, rather than restore traditional stability to explain the diversity, complexity, interconnectedness, and directional thrust of living organisms (295);
> *Dynamism.* They recognize that life is dynamic, developmental, and in a permanent process of emergence (295–96);
> *Interdependence.* Biological models designed to explain all these different aspects of interdependence, co-existence, and mutual determination between discrete living beings may all have something to tell us about how human political communities

can persist, adapt, and even govern themselves without self-destructing (296).

Along with their positive additions to the creation of politics, the perceived dangers of biological metaphors in the social sciences must be addressed. For IR, attention is often focused on the dangers of biology itself through dual-use biological research on infectious diseases like smallpox and influenza, and more broadly, biological metaphors are often avoided because they seem to carry with them illiberal and conservative connotations. To add to this, biology can be used for the "strict determination and limitation of social roles" and proving societal and cultural opinions about difference, race, violence, gender, and so on, as "natural" and unavoidable constraints on politics.[7] But as I endeavor to demonstrate, modern biology and current biological thinking is no longer based on static equilibrium or a bare struggle for survival expounded by neo-Darwinist theories of evolution. Of course, an important point to remember is that even with the metaphorical potential of modern biology, it is the case that "analogy is not homology."[8] Any transfer from biology to politics should be examined critically.

The lively vessel and the immune system are re-presented as examples of a posthuman becoming that finds itself as one of many "modest examples of biocultural hope,"[9] where much of the theoretical focus has been on pessimism and thanatopolitics, or politics focused on the power of death, to explain our current moment rather than analyzing the power of life to confound sovereign control. Life, understood as an emergent property in a complex system, cannot be explained by the sum of its parts; in biological terms, "none of the component molecules of a cell are alive, only a whole cell lives."[10] If emergence is difficult to define in biological terms, the sovereign desire to control and capture emergence is apparent. This focus necessitates a return to the immune system as a complex system to find the posthuman potential.

Posthuman Becoming

As I demonstrate throughout the book, we have never been wholly human, biologically or otherwise, but the explicit move to the posthuman does need explanation. Must the culturally specific idea of the human be retained if we recognize the damage it has done? Bodies that do not match those of the Enlightenment project—namely, white, male, and of European descent—were historically, and continue to be, defined as sub- or nonhuman. Our humanity is created at the expense, and suffering, of all who are outside this definition. This complicates defining the posthuman: it is not simply a facile, temporal definition of the human that comes after now. Cary Wolfe insists that, much like Lyotard's discussion of the postmodern, the posthuman is both before and after humanism. To come before, posthumanism incorporates both biological embodiment and human coevolution with technology, tools, and external archives like language and culture and it "names a historical moment in which the decentering of the human by its imbrication in technical, medical, informatic, and economic networks is increasingly impossible to ignore."[11]

To understand posthumanism, humanism must also be placed historically. It encompasses a shift in beliefs about the location of moral agency. Contra pre-Enlightenment thinking grounded in the belief of a transcendental or religious basis for ethics and morals, humanists locate moral agency in human rationality. Universal beliefs come from our common experiences and the combined teachings of humankind rather than religious texts or belief in God or other transcendent beings. This was a revolutionary shift in ethical and moral foundations. In this sense, posthumanism does come after humanism as a specific historical phenomenon that "points toward the necessity of new theoretical paradigms . . . after the cultural repressions and fantasies, the philosophical protocols and evasions" of humanism.[12] Posthumanism will also be a radical shift to locate moral agency and ethics in wider earthly systems and not just human rationality, and specifically Western, European, masculinized rationality.

The concept of the posthuman aids the lively vessel and contaminated state in two crucial ways. The first, as discussed in the introduction, is methodological. It allows for the overlapping of things, or objects, and social concepts in such a way that the researcher can learn more about a set of relations than is possible by looking at each discretely. In the case of the body, as I have outlined in the preceding chapters, seeing the human body without its nonhuman coevolved symbionts does not offer the fullest picture for new ideas of the individual through the biological sciences.

Following from this, the second way the posthuman can be defined is as a creative and ontic refiguration of the individual in much the same way as I have done with the biological sciences and the body politic based on its enmeshment in planetary processes. Theorists use this term to explicitly complicate and even shed prior definitions of the human and to question what defines Homo sapiens as human. In short, it is an attempt to think differently about bodies and their relation to discourses of power. More broadly, the need for contextuality when discussing the posthuman is, as Wolfe writes, an opportunity rather than a "cautionary tale."[13] N. Katherine Hayles writes of the pleasure in thinking about the human in new ways: "the posthuman evokes the exhilarating prospect of getting out of some of the old boxes and opening up new ways of thinking about what being human means."[14]

That said, it is important not to forget the body's form in these refigurations. The human body, with its microbiome, retroviruses, and symbiotic partners is one such way to reimagine the body. It is equally important to remember that bodies need to be theorized with their plurality and experiences intact. To neglect this is to forget the politics and ideology that help to create the individual, the person, and the human in modernity. To forget will only serve to incorporate politics, in this case the liberal individual, as if it were natural fact.[15] This is compounded by theorists who draw on the posthuman and their tendency to focus on it as a way to reanimalize the human or reconnect with nature to realize and nourish our interconnected existence on the planet. The posthuman cannot be an easy acceptance of our animal nature and continued acceptance of the dichotomous thinking inherent in nature/culture and human/animal. Not only has centuries of philosophy relied on the alleged distinctions between the human animal and other animals, animality has been used to belittle, discount, distort, mangle, and slaughter those considered not human, including both human and nonhuman animals. Indeed, animalizing populations and individuals is one of the warning signs of a coming genocide. Excising the non-, sub-, and unhuman from the "human" has justified the oppression, domination, violence, and cruelty that lurk beneath the creation of liberal society and its laws.[16]

In other words, posthuman theories must measure the tangled relationship that people of color and women have had with patriarchal and racist systems of thought and institutions created from humanism. Having been treated as nonhuman, or less than human, understandably leaves those who have suffered wary of coming back to the animal as a positive, emancipatory project.[17] If we have yet to treat many as fully human, how are we to understand the posthuman? Rosi Braidotti, in the first pages of her book on the posthuman, stresses not only that it is difficult to say with certainty that we have always been human but also that "some of us are not even considered fully human now, let alone at previous moments of Western social, political, and scientific history."[18] This means, of course, that what we mean by "human" is socially, politically, and scientifically understood with a regime of rights built around it. In the Western tradition adopted by IR, this means the human is a citizen, a Cartesian and Kantian subject, a property holder, and most often European (or of European descent) and male.[19]

In addition to following technological, scientific, and medical challenges to the human, this means paying close attention to those discussions in which race and gender are not mentioned or in which the human body is spoken of as if it has not been constructed prior to its becoming "post" human. Such silences have the undesirable effect of assuming that race is a pregiven and essentialized part of the human, or that "bare life" is not always already part of a racialized assemblage that has defined certain bodies as more dangerous and killable than other bodies.[20] The humanist and state projects[21] of racism, misogyny, autarkic sovereignty, and power politics are also linked to human economic systems that destroy, despoil, and pillage natural resources and planetary systems for the good of the few and the immiseration of the many. One has only to read the early debates between Vittorio and Las Casas and the actions of the colonizers in the Americas to know the outcomes of this kind of humanism in the New World.[22]

The posthuman, if it remembers its history, explodes these ideas of the human and it "introduces a qualitative shift in our thinking about what exactly is the common reference for our species, our polity, and our relationship with other inhabitants of the planet."[23] Recognizing past exclusions from humanity can lead to fuller and more socially and ecologically minded accounts of the human animal. The posthuman questions the humanist's belief in rationality and reason as the basis for moral agency; reason, as it is used here, is a Western creation, and it is often deployed as a way to keep those who are not Western, or the Other, from attaining full rights. In other words, it is unacceptable to treat human animals and nonhuman animals as objects or resources to use and abuse. This will only reproduce past and present hierarchies and power structures that are harmful to all species on the planet. Lines of radical solidarity need to be created along all lines of difference.

As a posthuman figure, the lively vessel summons a different ontology that insists that a detached domination of nature is not the best way to proceed. It aids in reconnecting the human body, and the abstract state that is its grand metaphor, to its environment. Container metaphors presuppose that one could stand outside of the world or even be able to distinguish inside from outside at a specific place and time. This supports dualism as natural and unavoidable, both with the world and with the human and nonhuman. "Instead of seeing dualist attachment and domination as a move, a tactic, a ploy, and a very specific way of living in the flow of becoming, we tend to mistake it for the world itself."[24] This dualist detachment and domination of nature, Pickering and Guzik argue, has veiled that humans live in "the thick of things." This veil will need to be drawn back to see the symmetric process of "the becoming of the human and nonhuman" (8).

Specifically, metagenomics, as both a set of research techniques and the community of scientists who work in this discipline, is beginning to recognize that a detached domination over nature is not the best way of proceeding in the world. The Committee on Metagenomics writes of our dependence on and interdependence with microbes in complex communities. Microbes transform the biosphere through fermentation and chemical cycles, and the committee recognizes that humans are not the authors or viewers of these processes, but intimately interconnected to these cycles, processes, and microbes for survival. "Although we can't usually see them, microbes are essential for every part of human life—indeed all life on Earth. Every process in the biosphere is touched by the seemingly endless capacity of microbes to transform the world around them."[25] In this vision, the world is a joint project between the human and the nonhuman. This is more than realizing that humans inhabit the same world, it is recognizing that we are composing the world together.[26]

In October 2015, a multidisciplinary group of scientists published a call to the larger scientific and political communities to develop much-needed tools for expanding the study of microbiomes to reveal their global diversity.[27] The first of these would be the Unified Microbiome Initiative Consortium: public and private researchers and agencies brought together to study the activities of earth's microbial systems. Although conceived of through a U.S. initiative, the group realizes that "Earth's biome is not defined by national borders, and efforts to unlock its secrets should go global."[28] The transformative power of microbiomes is only now becoming clear and local solutions to global problems may be the way to fight the larger processes of climate change, but only if a global effort is coordinated.

A holistic understanding of the role of Earth's microbial community and its genome—its microbiome—in the biosphere and in human health is key to meeting many of the challenges that face humanity in the twenty-first century, from energy to infection to agriculture.[29]

Metaphorical frames built from these understandings of microbial communities aid in understanding system-based understandings of complex processes in the international realm. Many problems that the state in its current form has been unable to address—warming oceans, pandemics, climate change, flows of immigrants and migrants—may be easier to address if a rich, system-based understanding is brought to bear on them.

Posthuman Mingling

Biopolitics and immunity have a natural affinity. At a theoretical level, philosophers have interrogated the idea of immunity as a cultural paradigm and metaphor for modern politics. Immunity has also had profound effects on the growth of biopolitical regimes, and intimately intertwines violence into our conceptions of the body. Biopolitics is defined as how power transforms itself to govern not just individuals, but populations; it emerges when politics takes hold of life. Biopolitical regimes are built from this power harnessed from life and from the sovereign's desire to regulate and control life. Although it is impossible to pin a moment or a founding text on this new configuration of politics, it is helpful to begin from Michel Foucault's writing on sexuality. In the latter half of *A History of Sexuality: Volume 1,* Foucault writes that sovereign power has shifted from the power to take life and let live, to the power to make live and let die. In other words, the sovereign exercises power by the lives he can take and the bodies he can rend apart; the sovereign suppresses death and allows subjects to live. Foucault identified that we have entered a new regime of power that can now disallow life to the point of death. The sovereign can wage war for the defense of populations and foster some lives by letting others live and letting others die.[30] Zygmunt Bauman likens this power to tending a garden in which "weeds" are controlled and plucked for the good of all the plants.[31] It is seemingly innocuous, but it holds a terrible judgment about some as worthy of life and others not, right at the edge of political vision.

Another important change in modernity and its regimes of power is that the awareness of immunity—as defined by self-defense of an organism from outside dangers—becomes an important condition for society and a metaphorical frame. Politics have always been preoccupied with defending life, but this does not "detract from the fact that beginning from a certain moment that coincides exactly with the origins of modernity, such a self-defensive requirement was identified not only and simply as a given, but both as a problem and a strategic option."[32] Roberto Esposito continues on to contrast the idea of *communitas,* or the binding of members of a community to reciprocal obligation with *immunitas* as the condition of dispensation from this obligation (50). It is here that Esposito inserts the immunitary paradigm as the specifically modern characterization of a politics based on life. While the ancient world, both historically and conceptually, oriented itself toward biopolitics with slavery, the killing of prisoners of war, control of the family through the patriarch and patriarchal regimes, and Greek writings on eugenics, it directed these toward communitarian concerns. Esposito is quick to stress that it is not his intention to argue that the premodern epoch did not raise the question of immunity. One could claim, Esposito maintains, that the history of civilization bears out the fact that the first condition for society is a "defensive apparatus" to make certain it can protect itself (54). However, the modern is distinct in its movement toward individual self-preservation (52–54). This "individual" interest versus communitarian concern marks an important point in Esposito's theory of the immunitary paradigm: immunity and community are opposites. Alfred Tauber, an immunologist and historian of science, writes that, through Metchnikoff's theory, "immunology clearly aligned itself with the biology of individuals, and indeed, one might easily argue it became the science of individuality at the expense of community."[33]

This is twinned with sovereignty and the nation-state understanding immunity as defense, and placing sovereign power in the very body itself in a new way. Ed Cohen, in his book *A Body Worth Defending,* writes that immunity, as a legal and juridical term, was coupled with an idea of the "natural right" of self-defense through Thomas Hobbes and the English Civil War. Early biomedicine fused these "two incredibly difficult, powerful, and yet very different (if not incongruous) political ideas into one, creating 'immunity-as-defense.'"[34] This coupling of immunity as self-defense and as a natural right firmly places the body in a regime of biopower, defined as the attachment of life to politics, and justifies violence as the base of all human (and nonhuman) relations. "Immunity places war and the sovereign exception inside the human body and naturalizes defense and violence as the principal way humans interact with their environments" (6). This "biological hybrid" is then transplanted into the actual body (3). This relationship acknowledges defense as a natural capacity of living organisms and is "a scientific practice that profoundly transforms how we conceive and address both healing and illness" (3–4):

> Through this potent conceptual alchemy, biological immunity insinuates itself at and as the intersection of two disparate, if not opposed, ways of organizing human interactions: war and law. To the extent that law tries to preempt war's violence (albeit by mobilizing its own violence) and to the extent that declarations of war seek to define violence's legal extent (albeit within their own jurisdiction), the two seem counterpoised. Hence, not only do immunity as a legal exemption and immunity as defense not necessarily correspond; they do not even necessarily coexist. Strictly speaking, where immunity exists there is no need of defense and where defending occurs there is no immunity. (6)

Despite this oppressive historical connection between biopolitics and immunity, there is important potential within the processes of immunity itself to contribute to a positive understanding of both biopolitics and modern political processes. This can open a space to reformulate biopolitics more positively by affirming life as vital and relational rather than as a purely mechanical reaction against that which is Other. Immunology also takes steps toward undoing ideas of atomistic individuality, or selfhood. From the perspective of ecological immunology, "there can be no circumscribed, defined entity that is designated *the self*" that can be defended against intruders.[35] It is equally important to stay silent, or tolerant, in immune function, and for the organism "as it responds along a continuum of behaviors to adapt to the challenges it faces. Indeed, 'identity' is determined by particular context,"[36] rather than in the protection of a static self. In dynamic, nonlinear systems, dichotomous distinctions cannot survive the complex calculations of responses that are needed. In other words, the self emerges out a result of these dynamic processes, and the defense of the integrity of this selfhood from pathogens is only a secondary phenomena.[37]

These moments of emergence are difficult to nurture or even to understand in the first place. If the immune response and the above characterizations of life and its capture by politics were seen through a dynamic, open system of understanding, new conceptions and practices of social-political life might appear. The body is not a closed system with aggressive tendencies toward who, or what, was outside, but rather one that systematically mediates its relations with others in terms of community, rather than immunity. Put another way, this would shift the focus from excluding others to including them: this would mean living in coexistence rather than self-defense.[38] Cohen asks a series of questions based on this metaphorical reimagining of community over immunity and wonders what the world would be like if "commune systems" explained our relations with the world. The human experience would be different if we imagined coexistence rather than self-defense as the basis for our well-being. Challenges to our corporeal being would become an ethical and political challenge to connect with others.[39] This would likely make the world a profoundly different place in which we could no longer defer harms to others without paying a price in our own well-being and happiness.

New vocabularies and metaphors should be nurtured to bring about new and more ecologically sensitive and socially just ways of being and becoming in the world. New methodologies need to be able to capture complexity and apply to global processes. One key methodology can be drawn from systems biology, which illuminates the potential for a different model of IR through its understanding of the organism as a transformational and reactive system.

Lively Systems

Systems biology is the study of systems with biological components. Living systems are dynamic and complex, and a major goal of this discipline is to "learn how the concurrent reactions and interactions of the lower-scale components of a cell, organism or society generate emergent properties visible at higher scales and higher layers of reality."[40] In other words, organisms emerge as part of a collective that performs collective actions with other entities. "Biologic entities, like Russian Babushka [nesting] dolls, are formed by embedded interactions at various scales and various layers. Interactions at a lower scale emerge as objects expressing their own properties at a higher scale. Scaling is the key to emergence; emergent properties arise as new objects from one scale to the next."[41]

Resonance with global politics becomes apparent when understanding relies on being able to observe and manipulate the system at more than one scale. An observer must also be able to convey dynamics or changes in the system in a fluid manner in real time. Traditionally, a researcher would build a biological model to test the validity of an understanding, or to see how well the model accounts for known behaviors, or how accurately it predicts changes in the system; but, crucially, systems biology turns the process of model making around: "first make a dynamic model that integrates the data, then you will understand. A model that represents faithfully the dynamic crossing of scales and layers is itself an explanation of the living system's emergent properties."[42] In this system, explanation always entails a combination of reduction (at a lower scale, for example, the cell) and extension (at a higher scale, for example, the organism) to come close to catching a dynamic system where emergence of life is the main concern. This is not to say that reductionism is not useful; living systems must be "carved into their component parts," but this carving will not have, by itself, "uncovered the fundamental laws from which we can deduce how a particular living system actually works."[43]

To make the analogical connections for International Relations, this would mean, quite simply, that the parts cannot make the whole if the observer desires to understand emergent properties in a system. The behavior of the system is not expressed by any of the lower scale components. As explained earlier in Latourian terms, following the transitions and interactions at different scales will allow for a fuller picture of the system. A state, like a cell, would emerge from its interactions with other states and agents. Interactions at one scale (the state) create objects at a higher scale (the international). This aids in seeing the state less as static object, and more as one that emerges from multiple actions and interactions.

The state has metaphorical actancy and emergent properties when considered through its lively internal messmates and tangled, yet productive, connections to outside itself. This, in turn, encourages new disciplinary understandings to rise in response to the dynamic models built. Progress in the social sciences and the natural sciences will be driven by new questions that demand new ways of thinking.[44] The complexity of the planet and the political issues that arise globally are coupled with the sheer amount of knowledge available to any one scholar. This demands organization that is best defined as "antedisciplinary," in that it precedes the organization of new disciplines. Inter- or transdisciplinary people will *invent* new ways to look at the world rather than combining existing skills to solve predefined problems.[45]

Coda

New Metaphors for Global Living

> *I sing the body electric,*
> *The armies of those I love engirth me and I engirth them,*
> *They will not let me off till I go with them, respond to them,*
> *And discorrupt them, and charge them full with the charge of the soul.*
>
> —Walt Whitman, *Leaves of Grass*

This book begins with a call for an expanded vocabulary for International Relations (IR): new words for a world facing novel challenges. The four chapters build an idea of connectedness between (and within) diverse bodies that allows for a reassessment of the core ontological, epistemological, ethical, and normative claims central to the figure of the body politic. It does this by mining the hermeneutic potential of the life sciences, specifically human microbiomes and immune systems. These potentials are refigured as the lively vessel and the contaminated state.

My broad aims with the examples are twofold. The first is to complicate the human subject as a body among other bodies and to demonstrate how the human body is connected and tangled with the material world and thereby enriches the way we shape politics and engage with the planet. The second brings a vocabulary and methodology to IR that expands notions of agency and participation. This allows us to interrogate global political conditions taking place in a complex array of interdependent and enmeshed spaces.

Bacteroides thetaiotaomicron teaches us that human and bacterial agency are part of an assemblage of multiple actors, and this calls attention to the nonhuman beings that aid in keeping the human body and its biosphere alive. As a step toward complicating IR's understanding of itself and its disciplinary subjects, the book brings the latter half of its title to bear on the discussion: "relations" trumped the "international" through the body politic as a nested set of permeable bodies rather than hardshelled nation-states competing in anarchical conditions ruled by fear and exclusion.

Put another way, my argument relies on IR's widening its scope and taking more "agents" into its disciplinary sights, but many of these agents will not be actors as we have previously understood them. In fact, *The Microbial State* presses IR, as a discipline, to take agent-centered theory more seriously—even more seriously than it has been asked to do previously. Beyond the sovereign state, agents may be invisible, like a virus, or large and complex, like the biosphere, and IR needs new words and conceptual apparatuses to engage and understand agents that have no "will" as we would define it. This is nonvolitional agency that is hard to locate within the anthropocentric agent-structure paradigm.[1]

Moreover, the guiding metaphors of IR need to be revealed and discussed, the most insidious being the reliance on the agent-structure metaphor. This metaphor not only mistakenly treats structures like the state and anarchy as if they were always already present, thereby forgetting their ideational birth, but this dichotomy is unable to understand change and transformation that does not match the traditional state and level of analysis understandings. To be clear, this book desires to move away from narrow methodological arguments about international change and continuity based solely on human concerns. IR's concerns are no longer the end of the Cold War, and how to best explain it through the thin metaphors of change irrupting from a loss of balance between two great powers or international anarchy. IR needs to be pulled into a new argument around systemic change and global-level transformation. Who and what are counted as agents are both bigger and more fine-grained than in previous theories of IR.[2]

Bruno Latour's "actant" captures an important piece of a missing story about our wider understanding of the world. In many ways, this should not be alien to International Relations; it needs only take the fiction of the state as a person more seriously. To answer Alexander Wendt's concern about whether or not the state is a person, I can respond with an affirmative: yes, the state is a person, but its body is like none we have discussed in IR previously. It is an actant in the Latourian sense. More than a construction of our ideas alone, it is a series of connections, translations, and matters of concern that are more, and less, than the sum of its parts. Interchange, complexity, and heterogeneity must be acknowledged as a political and ethical narrative in a world of transnational flows. It us not just that we have bodies, both real and imagined, that experience the world, but that the world acts upon us, too. This commitment to transnationality, or transversal flows, exceeds and challenges movements of capital or intellectual property as controlled by global capitalism. In fact, the communities built from lively vessels do not necessarily owe fealty to the state, or, if they do, the state is a hybrid form of nested bodies relying on each other in newly recognized and radical ways. Simply put, a single political institution cannot handle the needs of the planet.[3] Different understandings of political life will need to be written, sketched, performed, and composed—poems of the planet contrasted with writs from the state.

Through metagenomics, parasitism, and emerging understandings of the human body in scientific and medical discourse, I argue that difference is the condition of possibility for the production of new subjectivities. These new productive forms, or the metaphors of lively vessels and contaminated states, create a new perspective for collective living. The divisions between inside and outside, and their successive reformulations—the state subject and the individual subject—do not have the ability to fully explain the relationships found between microbiomes/humans and hosts/symbionts. This new idea of the body politic can reconstruct a concept of the common that is a "collection of singularities, a cooperative fabric that links together infinite singular activities,"[4] rather than the dichotomous violence inherent in the logic of the modern state.

Metaphors allow us, as the author and the reader, to reflect on the tools and framings needed to think beyond current and historical forms and to refigure the body politic from a different vantage point. This offers us a way to imagine, and dream about, the body in such a way that it opens new horizons for political organization and hopes to reveal our presuppositions about the body and its connection to politics and how they are culturally and scientifically produced in particular ways, rather than natural, unassailable, and untouched by politics and ideological production. My suggestive metaphors describe political bodies without devaluing bodily life and experience. This aims, as Nigel Thrift writes, to cultivate power through imagination, "where imagination is understood as the ability to express possible/play/pretend beliefs and emotions that might become the basis of a better world."[5] Bacteria, nations, parasites, guts, and bodies thrive as lively vessels to better capture the flows, immersions, circuits, and heterogeneities between and among a plurality of actors. These are new models of affectivity, or "spaces of speculation" to provoke and invoke new forms of intelligibility in politics and social life (141).

For these models and spaces, I choose scientific views that remain on the outside, or the fringe, of mainstream thought.[6] In their respective fields, scientists are involved in their own debates over the constitution of the world and how we can best represent that world through language, questioning the very same configurations of identity/difference, self/other, and human/nonhuman, and analyzing the value (and dangers) of scientific discoveries to social understandings.

These alternate visions of science demonstrate that hybridity and heterogeneity are necessary elements for understanding the nature of the individual and the individual's connection to the larger world. I do not want to forward science as a Western imperial project, and this desire is evidenced by the examples that I choose to analyze. In part, the language of science is used to highlight and complicate the easy reliance on earlier and outdated conceptions of science, and its epistemological role in IR, rather than introduced as a way that makes IR an "equally scientific" discipline by referencing science and biology.

The Microbial State articulates a sincere desire to redesign, traverse, and complicate the body politic; it wants to breathe new life into a philosophical construct and fill it with fantastical forms and unlikely beasts. As Nietzsche reminds us, teaching abstract truths means first enticing the senses. The future that lies ahead in this geologic epoch now called the Anthropocene demands that we reassess the systems and forms of knowledge that have guided the human race, and with us all other species on earth, to the brink of, at best, severe ecological change, and at worst, catastrophe for our species and countless others. Our concepts may need to be as messy and complex as our bodies to understand and survive in the times ahead.

I leave the last word to the poet who began each chapter. From the large to the small, and how we might understand ourselves as humans through the earth's body as metaphor, Whitman's words guide us through these subtle analogies and how we might live, in both time and space, together:

> Who having considered the body finds all its organs and parts good,
> Who, out of the theory of the earth and of his or her body understands by subtle analogies all other theories,
> The theory of a city, a poem, and of the large politics of these States;

Who believes not only in our globe and moon, but in other globes and moons,
Who, constructing the house of himself or herself, not for a day but for all time, sees races, eras, dates, generations,
The past, the future, dwelling there, like space, inseparable together.[7]

Notes

Introduction

1. Paul J. Crutzen, "Geology of Mankind," *Nature* 415 (2002): 23.

2. Mel Y. Chen, *Animacies: Biopolitics, Racial Mattering, and Queer Affect* (London: Duke University Press, 2012); Naomi Klein, *This Changes Everything: Capitalism vs. the Climate* (London: Penguin, 2014); Roy Scranton, *Learning to Die in the Anthropocene* (San Francisco: City Light Books, 2015); Anna Lowenhaupt Tsing, *The Mushroom at the End of the World: On the Possibility of Life in the Capitalist Ruins* (Princeton, N.J.: Princeton University Press, 2015); McKenzie Wark, *Molecular Red: Theory for the Anthropocene* (London: Verso Books, 2015).

3. Lesley Instone and Affrica Taylor, "Thinking about Inheritance through the Figure of the Anthropocene, from the Antipodes and in the Presence of Others," *Environmental Humanities* 7 (2015): 133–50.

4. N. Katherine Hayles, *How We Became Posthuman: Virtual Bodies in Cybernetics, Literature, and Informatics* (Chicago: University of Chicago Press, 1999), 4.

5. Sylvia Wynter, "Unsettling the Coloniality of Being/Power/Truth/Freedom: Towards the Human, after Man, and Its Overrepresentation—An Argument," *New Centennial Review* 3, no. 3 (Fall 2003): 260.

6. Alexander Weheliye, *Racializing Assemblages, Biopolitics, and Black Feminist Theories of the Human* (Durham, N.C.: Duke University Press, 2014).

7. Charles Mills, *The Racial Contract* (Ithaca, N.Y.: Cornell University Press, 1997); Robert Vitalis, "The Graceful and Generous Liberal Gesture: Making Racism Invisible in American International Relations," *Millennium: Journal of International Studies* 29, no. 2 (2000): 331–56; Robert Vitalis, *White World Order, Black Power Politics: The Birth of American International Relations* (Ithaca, N.Y.: Cornell University Press, 2015).

8. John Heskett, *Design: A Very Short Introduction* (Oxford: Oxford University Press, 2002), 11–12.

9. Ibid., 12.

10. Benedict Anderson, *Imagined Communities: Reflections on the Origin and Spread of Nationalism* (London: Verso Books, 1991).

11. Thanks to Amy Groleau for this simple formulation.

12. Giorgio Agamben, *Homo Sacer: Sovereign Power and Bare Life* (Stanford, Calif.: Stanford University Press, 1998).

13. Ed Yong, *I Contain Multitudes: The Microbes within Us and a Grander View of Life* (New York: HarperCollins, 2016).

14. Nicholas Gane, "When We Have Never Been Human, What Is to Be Done? Interview with Donna Haraway," *Theory, Culture and Society* 23 (2006): 135–58.

15. Jane Bennett, *Vibrant Matter: A Political Ecology of Things* (Durham, N.C.: Duke University Press, 2010).

16. Donna J. Haraway, *Staying with the Trouble: Making Kin in the Chthulucene* (Durham, N.C.: Duke University Press, 2016).

17. Andrew Pickering, "Asian Eels and Global Warming: A Posthumanist Perspective on Science and the Environment," *Ethics and Environment* 10, no. 2 (2005): 29–43.

18. Ed Cohen, "Human Tendencies," *e-misférica* 10, no. 1 (Winter 2013), accessed January 25, 2017, http://hemisphericinstitute.org/hemi/en/e-misferica-101/cohen.

19. Ibid.

20. Francis Fukuyama, *The End of History and the Last Man* (New York: Avon Books, 1992).

21. Editorial, "The Nation State Is Dead. Long Live the Nation-State," *Economist* (December 1995–January 1996): 15–18.

22. Patrick Thaddeus Jackson, "Forum Introduction: Is the State a Person? Why Should We Care?," *Review of International Studies* 30 (2004): 255–58.

23. R. B. J. Walker, *After the Globe, Before the World* (London: Routledge, 2010), 130.

24. Stefanie Fishel, "Theorizing Violence in the Responsibility to Protect," *Critical Studies on Security* 1, no. 2 (2013): 204–18.

25. Robert H. Blank and Samuel M. Hines Jr., *Biology and Political Science* (New York: Routledge, 2002), 1.

26. For an example of a Darwinian approach to IR, see Bradley A. Thayer, *Darwin and International Relations* (Lexington: University Press of Kentucky, 2004). Whereas the book is of interest to sociobiologists, it is not intended to be a comparative study of the similarities between bacterial communities and human communities. E. O. Wilson's work stands out as the best example in the genre, see *Sociobiology* (Cambridge, Mass.: Belknap Press of Harvard University Press, 1980). For a contribution to the debate between genetic and environmental determinism and its influence on human behavior through a study on bacterial quorum sensing and biofilm, see Matthew R. Parsek and E. P. Greenberg's article, "Sociomicrobiology: The Connection between Quorum Sensing and the Biofilm," *Trends in Microbiology* 13, no. 1 (2005): 27–33. This book draws more upon Henri Bergson and A. N. Whitehead as well as Jakob Von Uexkull and his studies on ethology and perception.

27. Arthur Lupia, *Genes, Cognition, and Social Behavior: Next Steps for Foundations and Researchers* (Washington, D.C.: National Science Foundation, 2001), 2.

28. William Connolly, *A World of Becoming* (Durham, N.C.: Duke University Press, 2011), 17.

29. Lupia, *Genes, Cognition, and Social Behavior*, 3.

30. Immaculada De Melo-Martin, "Creating Reflective Spaces: Interactions between Philosophers and Biomedical Scientists," *Perspectives in Biology and Medicine* 52, no. 1 (2009): 39–47.

31. Connolly, *A World of Becoming*, 17, 21.

32. Daniel Gibson et al., "Creation of a Bacterial Cell Controlled by a Chemically Synthesized Genome," *Science* 329 (2010): 52–56.

33. Michel Foucault, "Truth and Power," in *Power/Knowledge: Selected Interviews and Other Writings, 1972–1977*, ed. Colin Gordon (New York: Pantheon Books, 1980), 112.

34. Bruno Latour, "Science Wars: A Dialogue," *Common Knowledge* 8, no. 1 (2002): 77.

35. Laurence Whitehead, "Enlivening the Concept of Democratization: The Biological Metaphor," *Perspectives on Politics* 9, no. 2 (2011): 291.

36. Timothy Morton, *The Ecological Thought* (Cambridge, Mass.: Harvard University Press, 2010).

37. Elling Ulvestad, *Defending Life: The Nature of Host–Parasite Relations* (Dordrecht: Springer, 2007), xii.

38. Paul Feyerabend, *Against Method* (London: Verso Books, 2010).

39. Diana Coole and Samantha Frost, eds., *New Materialisms: Ontology, Agency, and Politics* (Durham, N.C.: Duke University Press, 2010), 4.

40. Cynthia Weber, *International Relations Theory: A Critical Introduction* (New York: Routledge, 2010).

41. Terrell Carver and Jernej Pikalo, *Political Language and Metaphor: Interpreting and Changing the World* (London: Routledge, 2008).

42. This project also builds on work about global politics and political metaphor. See Michael P. Marks, *The Prison as Metaphor: Re-Imagining International Relations* (New York: Peter Lang, 2003); Michael P. Marks, *Metaphors in International Relations Theory* (New York: Palgrave Macmillan, 2011); Richard Little, *The Balance of Power in International Relations: Metaphors, Myths and Models* (Cambridge: Cambridge University Press, 2007); *The Body in Human Inquiry*, ed. Vicente Berdayes, Luigi Esposito, and John W. Murphy (Cresskill: Hampton Press, 2004); Edward Slingerland, Eric M. Blanchard, and Lyn Boyd-Johnson, "Collision with China: Conceptual Metaphor Analysis, Somatic Marking, and the EP-3 Incident," *International Studies Quarterly* 51 (2007): 53–57; Markus Kornprobst, Nisha Shah Vincent, and Ruben Zaiotti, eds., *Metaphors of Globalization: Mirrors, Magicians and Mutinies* (New York: Palgrave Macmillan, 2008); Francis

A. Beer and Christ'l De Landtsheer, eds., *Metaphorical World Politics* (East Lansing: Michigan State University Press, 2004); Simon Dalby, "Ecological Metaphors of Security: World Politics in the Biosphere," *Alternatives* 23, no. 3 (1998): 291–320.

43. George Lakoff and Mark Johnson, *Metaphors We Live By* (Chicago: University of Chicago Press, 1980), 230.

44. Naomi Quinn, "The Cultural Basis of Metaphor," in *Beyond Metaphor: The Theory of Tropes in Anthropology*, ed. James W. Fernandez (Stanford, Calif.: Stanford University Press, 1991), 57.

45. Jose Julian Lopez, "Notes on Metaphors, Notes as Metaphors: The Genome as Musical Spectacle," *Science Communication* 29, no. 7 (2007): 7–34.

46. Lakoff and Johnson, *Metaphors*, 25.

47. Milan Kundera, *The Unbearable Lightness of Being* (New York: HarperCollins, 1984), 11.

48. Keith L. Shimko, "The Power of Metaphor and Metaphors of Power," in *Metaphorical World Politics*, ed. Francis A. Beer and Christ'l De Landtsheer (East Lansing: University of Michigan Press, 2004), 202.

49. George Lakoff, "Metaphor and War: The Metaphor System Used to Justify War in the Gulf," *Viet Nam Generation Journal* 3 (1991), accessed January 25, 2017, http://www2.iath.virginia.edu/sixties/HTML_docs/Texts/Scholarly/Lakoff_Gulf_Metaphor_1.html.

50. Quoted in Shimko, "The Power of Metaphor and Metaphors of Power," 202.

51. Friedrich Nietzsche, *The Portable Nietzsche*, ed. and trans. Walter Kaufman (New York: Viking Press, 1954), 46.

52. Giovanna Borradori, ed., *Philosophy in a Time of Terror: Dialogues with Jürgen Habermas and Jacques Derrida* (Chicago: University of Chicago Press, 2003).

53. Walt Whitman, "Collect: Notes Left Over," in *The Whitman Reader*, ed. Maxwell Geismar (New York: Pocket Books, 1955), 456.

54. Michael Moon, *Disseminating Whitman: Revision and Corporeality in Leaves of Grass* (Cambridge, Mass.: Harvard University Press, 1991), 6.

55. Peter Simonson, "A Rhetoric for Polytheistic Democracy: Walt Whitman's 'Poem of Many in One,'" *Philosophy and Rhetoric* 36, no. 4 (2003): 353.

56. Hannah Arendt, *The Human Condition* (Chicago: University of Chicago Press, 1958), 8.

57. Simonson, "A Rhetoric for Polytheistic Democracy," 355.

58. Rosi Braidotti, *Transpositions: On Nomadic Ethics* (Cambridge: Polity Press, 2006), 265.

59. Jussi Parikka, *Insect Media: An Archaeology of Animals and Technology* (Minneapolis: University of Minnesota Press, 2010), xxiii.

60. Braidotti, *Transpositions*, 270.

61. Bruno Latour, *Reassembling the Social: An Introduction to Actor-Network-Theory* (Oxford: Oxford University Press, 2005), 260.

62. Ibid., 253.

63. Arendt, *The Human Condition*, 2.

64. William Connolly, "Materialities of Experience," in *New Materialisms: Ontology, Agency, and Politics*, ed. Diana Coole and Samantha Frost (Durham, N.C.: Duke University Press, 2010), 196.

65. William Connolly, *Pluralism* (Durham, N.C.: Duke University Press, 2005).

66. Connolly, "Materialities of Experience," 195.

67. Bennett, *Vibrant Matter*, 10.

68. Gane, "We Have Never Been Human," 145.

69. Lynn Margulis and Dorion Sagan, *Acquiring Genomes: A Theory of the Origin of the Species* (New York: Basic Books, 2002), 19.

70. Coole and Frost, *New Materialisms*, 4.

1. Corporeal Politics

1. For notable exceptions, see Diana Saco, "Gendering Sovereignty: Marriage and International Relations," *European Journal of International Relations* 3 (1997): 291–318; Cynthia Weber, "Performative States," *Millennium: Journal of International Studies* 27 (1998): 77–95; Iver B. Neumann, "The Body of the Diplomat," *European Journal of International Relations* 14 (2008): 671–95; Rosemary E. Shinko, "Ethics after Liberalism: Why Autonomous Bodies Matter," *Millennium: Journal of International Studies* 38 (2010): 723–45; Lauren B. Wilcox, *Bodies of Violence: Theorizing Embodied Subjects in International Relations* (Oxford: Oxford University Press, 2014); and Jessica Auchter, *The Politics of Haunting and Memory in International Relations* (New York: Routledge, 2014).

2. Allegedly said by General Tommy Franks at Bagram Air Force Base.

3. Diana Coole and Samantha Frost, eds., *New Materialisms: Ontology, Agency, and Politics* (Durham, N.C.: Duke University Press, 2010), 19.

4. Thomas Hobbes, *Leviathan: An Authoritative Text*, ed. Richard Flathman and David Johnston (New York: Norton, 1997), 4.

5. Michel Foucault, *Power/Knowledge: Selected Interviews and Other Writings, 1972–1977*, ed. Colin Gordon (New York: Pantheon Books, 1980), 58.

6. Jean-Luc Nancy, *Corpus* (New York: Fordham University Press, 2006), 155.

7. The four Great Debates are: realist–idealist in the 1930s and 1940s in response to World War II and Nazi Germany; realist versus the behavioralists around positivist scientific approaches; the inter-paradigm debate between realism, liberalism, and critical approaches; and the positivist/post-positivist divide between the so-called rationalists and reflectivists.

8. Thanks to Stacey Yadav-Philbrook for this idea about the "un-stated" state. She introduced another productive metaphorical play on words that aids in organizing this book.

9. Bruno Latour, "From Real Politik to Dingpolitik: Or, How to Make Things Public," in *Making Things Public*, ed. Bruno Latour and Peter Weibel (Cambridge, Mass.: MIT Press, 2005), 6.

10. Ibid., 6.

11. Thomas Hobbes, *Leviathan: An Authoritative Text,* ed. Richard Flathman and David Johnston (New York: Norton, 1997).

12. John O'Neill, *Five Bodies: Refiguring Relationships* (London: Sage, 2004).

13. Anthony Burke, *Beyond Security, Ethics and Violence: The War against the Other* (New York: Routledge, 2007), 36.

14. Steven Shapin and Simon Schaffer, *Leviathan and the Air Pump: Boyle and the Experimental Life* (Princeton, N.J.: Princeton University Press, 1985).

15. Juhanna Lemetti, *Historical Dictionary of Hobbes's Philosophy* (Lanham, Md.: Scarecrow Press, 2012), 161.

16. Foucault, *Power/Knowledge,* 133.

17. Anne Tickner, "Dealing with Difference: Problems and Possibilities for Dialogue in International Relations," *Millennium: Journal of International Studies* 39, no. 3 (2001): 611.

18. Robert Cox, "Social Forces, States, and World Orders: Beyond International Relations Theory," *Millennium: Journal of International Studies* 10, no. 2 (1981): 128.

19. Foucault, *Power/Knowledge,* 131.

20. Gilles Deleuze, *Empiricism and Subjectivity: An Essay on Hume's Theory of Human Nature* (New York: Columbia University Press, 1991), 106.

21. Jens Bartelsen, *A Genealogy of Sovereignty* (Cambridge: Cambridge University Press, 1995), 137.

22. Timothy Morton, *Ecology without Nature: Rethinking Environmental Aesthetics* (Cambridge, Mass.: Harvard University Press, 2007), 26–27.

23. Bruno Latour, "Science Wars: A Dialogue," *Common Knowledge* 8, no. 1 (2002): 78.

24. Bruno Latour, *Pandora's Hope: Essays on the Reality of Science Studies* (Cambridge, Mass.: Harvard University Press, 1999), 15.

25. Christian Büger and Frank Gadinger, "Reassembling and Dissecting: International Relations from a Science Studies Perspective," *International Studies Perspective* 8 (2007): 90–110.

26. Bruno Latour, *Reassembling the Social: An Introduction to Actor-Network-Theory* (Oxford: Oxford University Press, 2005), 187.

27. Ibid., 187.

28. Paul Feyerabend, *Against Method* (London: Verso Books, 2010), 15.

29. See Jens Bartelsen, *Critique of the State* (Cambridge: Cambridge University Press, 2001).

30. Feyerabend, *Against Method*, 15.

31. William Connolly, "Then and Now: Participant-Observation in Political Theory," in *Oxford Handbook of Political Theory*, ed. John S. Dryzek, Bonnie Honig, and Anne Phillips (Oxford: Oxford University Press, 2006), 838.

32. Ibid.

33. Donna Haraway, "The Promises of Monsters: A Regenerative Politics for Inappropriate Others," in *Cultural Studies* (New York: Routledge, 1992), 296.

34. Tickner, "Dealing with Difference," 617.

35. For example, Jürgen Habermas, in *The Future of Human Nature* (Cambridge: Polity Press, 2003), argues for a morality that is embedded in an "ethics of the species." Animals, he writes, benefit from our moral duties to other humans, but "do not belong to a universe of members who address intersubjectively accepted rules and orders *to one another*" (33). Our ability to make rules and norms and to address each other through language (deliberative democracy) is the basis for creating community and "human dignity." This leaves any other life form or lifeworld just a matter of regulation and responsibility rather than something that exerts any force on politics. Habermas sees any questioning of human nature, or acceptance of any nonhuman as he defines it, as "bought at the price of a *reductionist* definition of humanity and morality" (ibid.).

36. Melissa Gregg and Gregory J. Seigworth, "An Inventory of Shimmers," in *The Affect Theory Reader*, ed. Gregg and Seigworth (Durham, N.C.: Duke University Press, 2010), 2.

37. Elizabeth Grosz, *Chaos, Territory, Art: Deleuze and the Framing of the World* (New York: Columbia University Press, 2008), 22.

38. Jane Bennett, *Vibrant Matter: A Political Ecology of Things* (Durham, N.C.: Duke University Press, 2010), 117.

39. Gilles Deleuze and Felix Guattari, *A Thousand Plateaus: Capitalism and Schizophrenia* (Minneapolis: University of Minnesota Press, 1987), 260.

40. Benedict de Spinoza, *The Ethics* (New York: Dover, 1955), 93.

41. Coole and Frost, *New Materialisms*, 245.

42. Hannah Arendt, *The Human Condition* (Chicago: University of Chicago Press, 1958), 2.

43. Foucault, *Power/Knowledge*, 114.

44. Spinoza, *The Ethics*, 132.

45. Mark Johnson, *The Meaning of the Body: Aesthetics of Human Understanding* (Chicago: University of Chicago Press, 2007), 276.

46. Lisa Blackman, *The Body* (Oxford: Berg, 2008), 15.

47. Foucault, *Power/Knowledge*, 58.

48. Silke Schicktanz, "Why the Way We Consider the Body Matters," *Philosophy, Ethics and Humanities in Medicine* 2, no. 30 (2007): 5.

49. Roberto Esposito, *Bios: Biopolitics and Philosophy* (Minneapolis: University of Minnesota Press, 2008), 84.

50. For a history of the monarchical body politic, see Ernst Kantorowicz, *The King's Two Bodies* (Princeton, N.J.: Princeton University Press, 1997).

51. Foucault, *Power/Knowledge*, 53.

52. Chris Shilling, *The Body and Social Theory* (London: Sage, 2003), 3.

53. Coole and Frost, *New Materialisms*, 19.

54. Johnson, *The Meaning of the Body*, 265.

55. Coole and Frost, *New Materialisms*, 19.

56. Shilling, *The Body and Social Theory*, vii.

57. Ibid., 12.

58. Ibid., 182.

59. Bruce Braun and Sarah J. Whatmore, eds., *Political Matter: Technoscience, Democracy, and Political Life* (Minneapolis: University of Minnesota Press, 2010), x.

60. William Connolly, *A World of Becoming* (Durham, N.C.: Duke University Press, 2011), 22.

61. Haraway, "The Promises of Monsters," 311.

62. Braun and Whatmore, *Political Matter*, x.

63. For a notable exception, see Büger and Gadinger, "Reassembling and Dissecting."

64. Ole Wæver, "A Not So International Discipline," in *Exploration and Contestation in the Study of World Politics*, ed. Peter J. Katzenstein, Robert Owen Keohane, and Stephen D. Krasner (Cambridge, Mass.: MIT Press, 1999), 51.

65. Latour, *Reassembling the Social*, 43.

66. Joost Van Loon, *Risk and Technological Culture: Towards a Society of Virulence* (London: Routledge, 2002), 11.

67. Latour, *Pandora's Hope*, 303.

68. Graham Harman, *Prince of Networks: Bruno Latour and Metaphysics* (Melbourne: re.press, 2009), 5.

69. Latour, *Pandora's Hope*, 308.

70. Bennett, *Vibrant Matter*, viii.

71. Van Loon, *Risk and Technological Culture*, 11.

72. Harman, *Prince of Networks*, 14.

73. Latour, *Pandora's Hope*, 311.

74. Latour, *Reassembling the Social*, 170.

75. Paul Thibodeau and Lera Boroditsky, "Metaphors We Think With: The Role of Metaphor in Reasoning," *PLoS ONE* 6, no. 2: 1–11.

76. George Lakoff and Mark Turner, *More Than Cool Reason: A Field Guide to Poetic Metaphor* (Chicago: University of Chicago Press, 1989), 58.

77. Bennett, *Vibrant Matter*, 120.

78. Latour, *Pandora's Hope*, 9.

79. Francis A. Beer and Christ'l De Landtsheer, eds., *Metaphorical World Politics* (East Lansing: Michigan State University Press, 2004), 16.

80. Ibid.

81. Richard Little, *The Balance of Power in International Relations: Metaphors, Myths and Models* (Cambridge: Cambridge University Press, 2007), 19; Kenneth Waltz, *Theories of International Politics* (Long Grove: Waveland Press, Inc., 1977); John Mearsheimer, *The Tragedy of Great Power Politics* (New York: W.W. Norton and Company, 2001); Hedley Bull, *The Anarchical Society: A Study of Order in World Politics* (New York: Columbia University Press, 1977).

82. Ibid., 5.

83. Karl W. Deutsch, *The Nerves of Government: Models of Political Communications and Control* (London: Free Press of Glencoe, 1963), 26.

84. George Lakoff, "Metaphor and War: The Metaphor System Used to Justify War in the Gulf," *Viet Nam Generation Journal* 3 (1991), http://www2.iath.virginia.edu/sixties/HTML_docs/Texts/Scholarly/Lakoff_Gulf_Metaphor_1.html.

85. Melinda Wenner, "The War against War Metaphors: The Age-Old Practice May Harm Both Science and Scientists," *Scientist*, February 16, 2007, accessed February 1, 2017, http://www.the-scientist.com/?articles.view/articleNo/24756/title/The-war-against-war-metaphors/.

86. Guardian Blog, "Libya Air Strikes-Monday 21 March Part 1," *Guardian*, March 2011, accessed February 1, 2017, https://www.theguardian.com/world/blog/2011/mar/21/libya-military-action-live-updates.

87. Roner Sofi, "MK Ben-Ari: Eradicate Treacherous Leftists," *Israel News*, January 5, 2011.

88. Mark Franchetti, "US Marines Turn Fire on Civilians at the Bridge of Death," *Sunday Times*, March 30, 2003.

89. Laurence Whitehead, "Enlivening the Concept of Democratization: The Biological Metaphor," *Perspectives on Politics* 9, no. 2 (2011): 293.

90. Lakoff, "Metaphor and War," accessed February 1, 2017, http://www2.iath.virginia.edu/sixties/HTML_docs/Texts/Scholarly/Lakoff_Gulf_Metaphor_1.html.

91. George Lakoff and Mark Johnson, *Metaphors We Live By* (Chicago: University of Chicago Press, 1980).

92. Anthony Giddens, *The Nation-State and Violence* (Berkeley: University of California Press, 1987), 120.

93. Georges Canguilhem, "Monstrosity and the Monstrous," in *The Body: A Reader*, ed. Mariam Fraser and Monica Greco (New York: Routledge, 2005), 12.

94. Lakoff, "Metaphor and War," accessed February 1, 2017, http://www2.iath.virginia.edu/sixties/HTML_docs/Texts/Scholarly/Lakoff_Gulf_Metaphor_1.html.

95. Little, *The Balance of Power in International Relations*, 19.
96. Michael P. Marks, *The Prison as Metaphor: Re-Imagining International Relations* (New York: Peter Lang, 2003), 156.
97. Michael P. Marks, *Metaphors in International Relations Theory* (New York: Palgrave Macmillan, 2011), 4.
98. David J. Cisneros, "Contaminated Communities: The Metaphor of 'Immigrant as Pollutant' in Media Representations of Immigration," *Rhetoric and Public Affairs* 11, no. 4 (2008): 571.
99. R. B. J. Walker, *After the Globe, Before the World* (London: Routledge, 2010), 246.
100. Veronique Mottier, "Metaphors, Mini-Narratives and Foucauldian Discourse Theory," in *Political Language and Metaphor: Interpreting and Changing the World*, ed. Terrell Carver and Jernej Pikalo (London: Routledge, 2008), 192.
101. A. David Napier, *The Age of Immunology: Conceiving a Future in an Alienating World* (Chicago: University of Chicago Press, 2003), 69.
102. Thibodeau and Boroditsky, "Metaphors We Think With," 10.
103. Adam Gorlick, "Is Crime a Virus or a Beast? When Describing Crime, Stanford Study Shows the Word You Pick Can Frame the Debate on How to Fight It," accessed September 24, 2016, http://news.stanford.edu/news/2011/february/metaphors-crime-study-022311.html.
104. Marks, *The Prison as Metaphor*, 153.
105. Terrell Carver and Jernej Pikalo, *Political Language and Metaphor: Interpreting and Changing the World* (London: Routledge, 2008), 3.
106. Ibid., 4.

2. Lively Subjects, Bodies Politic

1. Myra J. Hird, *The Origins of Sociable Life: Evolution after Science Studies* (London: Palgrave Macmillan, 2009), 30.
2. Committee on Metagenomics, *The New Science of Metagenomics: Revealing the Secrets of Our Microbial Planet* (Washington, D.C.: National Academies Press, 2007), 18.
3. Carl Zimmer, *Microcosm* (New York: Pantheon Books, 2008).
4. Anne Maczulak, *Allies and Enemies: How the World Depends on Bacteria* (Upper Saddle River, N.J.: FT Press, 2010), 30.
5. See William H. McNeill, *Plagues and Peoples* (Garden City, N.Y.: Anchor Books, 1976).
6. David B. Dusenbery, *Living at Micro Scale: The Unexpected Physics of Being Small* (Cambridge, Mass.: Harvard University Press, 2009), 3.
7. Stewart Brand, "Microbes Run the World," accessed August 7, 2013, http://www.edge.org/q2011/q11_16.html#brand.

8. Committee on Metagenomics, *The New Science of Metagenomics*, 12.

9. Ibid., 2.

10. John Dupree and Maureen O'Malley, "Metagenomics and Biological Ontology," *Studies in History and Philosophy of Biological and Biomedical Sciences* 38 (2007): 837.

11. Committee on Metagenomics, *The New Science of Metagenomics*, 13.

12. Dupree and O'Malley, "Metagenomics and Biological Ontology," 835.

13. Committee on Metagenomics, *The New Science of Metagenomics*, 2.

14. Dupree and O'Malley, "Metagenomics and Biological Ontology," 835.

15. Committee on Metagenomics, *The New Science of Metagenomics*, 12.

16. National Public Radio, *Bacterial Bonanza: Microbes Keep Us Alive*, September 15, 2010, accessed February 2, 2017, http://www.npr.org/templates/story/story.php?storyId=129862107.

17. Katherine Harmon, "Genetics in the Gut," *Scientific American*, May 1, 2010, accessed February 2, 2017, https://www.scientificamerican.com/article/genetics-in-the-gut-may-2010/.

18. Ed Yong, "Introduction to the Microbiome," *Discovery Magazine*, accessed September 24, 2016, http://blogs.discovermagazine.com/notrocketscience/2010/08/08/an-introduction-to-the-microbiome/#.UTDGWxnlVZY.

19. Editorial, "A Universe of Us," *New York Times*, July 20, 2010.

20. Robert Martone, "The Neuroscience of the Gut," *Scientific American*, April 19, 2011.

21. Kate Becker, "Less Than One Percent Human," *Nova Inside*, accessed September 25, 2016, http://www.pbs.org/wgbh/nova/insidenova/2011/02/less-than-one-percent-human.html.

22. Brigitte Nerlich and Iina Hellsten, "Beyond the Human Genome: Microbes, Metaphors and What It Means to Be Human in an Interconnected Post-Genomic World," *New Genetics and Society* 28 (2009): 27–28.

23. Rosa Rhodes, "Human Microbiome Research and the Social Fabric," *Human Microbiome Project*, accessed September 24, 2016, http://hmpdacc.org/doc/d3.s7.t3%20-%20Rhodes%20-%20Social%20fabric%20implications.pdf.

24. Committee on Metagenomics, *The New Science of Metagenomics*, 30.

25. Eric T. Juengst, "Metagenomic Metaphors: New Images of the Human from a Translational Genomic Approach," in *New Visions of Nature: Complexity and Authenticity*, ed. Martin Drenthen, Jozef Keulartz, and James Proctor (London: Springer, 2009), 133.

26. Thomas S. Kuhn, *The Structure of Scientific Revolutions* (Chicago: The University of Chicago Press, 1970), 6.

27. Dupree and O'Malley, "Metagenomics and Biological Ontology," 842.

28. George Lakoff and Mark Johnson, *Metaphors We Live By* (Chicago: University of Chicago Press, 1980), 29.

29. A study into social and biological ideas of the skin itself would be a productive way to get at the theoretical and material issues at stake in making borders.

30. Lakoff and Johnson, *Metaphors We Live By*, 25.

31. Zimmer, *Microcosm*, 20.

32. Lynn Margulis and Dorion Sagan, *Acquiring Genomes: A Theory of the Origin of the Species* (New York: Basic Books, 2002), 19.

33. Elizabeth Grosz, *Chaos, Territory, Art: Deleuze and the Framing of the World* (New York: Columbia University Press, 2008), 80.

34. N. Katherine Hayles, *How We Became Posthuman: Virtual Bodies in Cybernetics, Literature, and Informatics* (Chicago: University of Chicago Press, 1999), 6.

35. Ibid., 60.

36. Grosz, *Chaos, Territory, Art*, 80.

37. Jussi Parikka, *Insect Media: An Archaeology of Animals and Technology* (Minneapolis: University of Minnesota Press, 2010), xxii.

38. Margulis and Sagan, *Acquiring Genomes*, 19.

39. Lisa Blackman, *The Body* (Oxford: Berg, 2008), 1.

40. Robert Krulwich, "Gut Bacteria Know Secrets about Your Future," accessed September 24, 2016, http://www.npr.org/blogs/krulwich.

41. Margulis and Sagan, *Acquiring Genomes*, 86.

42. Dupree and O'Malley, "Metagenomics and Biological Ontology," 840.

43. Margulis and Sagan, *Acquiring Genomes*, xii.

44. Dupree and O'Malley, "Metagenomics and Biological Ontology," 841.

45. See Gilles Deleuze, "Postscript on Control Societies," in *Negotiations, 1972–1990* (New York: Columbia University Press, 1995), 177–182.

46. Juengst, "Metagenomic Metaphors," 133.

47. Jane Bennett, *Vibrant Matter: A Political Ecology of Things* (Durham, N.C.: Duke University Press, 2010), 115–16.

48. See *Global State of the Oceans Report 2013*, http://www.stateoftheocean.org/research.cfm.

49. H. Steven Wiley, "If Bacteria Can Do It?," *Scientist—Magazine of the Life Sciences*, May 1, 2011, http://www.the-scientist.com/article/display/58131/#ixzz1PGQjGit5.

50. Bennett, *Vibrant Matter*, 115–16.

51. Michel Foucault, "The Subject and Power," *Critical Inquiry* 8, no. 4 (Summer 1982): 777–95.

52. Margulis and Sagan, *Acquiring Genomes*, 16.

53. Dupree and O'Malley, "Metagenomics and Biological Ontology," 842.

54. Ibid.

55. Juengst, "Metagenomic Metaphors," 141.

56. A. David Napier, *The Age of Immunology: Conceiving a Future in an Alienating World* (Chicago: University of Chicago Press, 2003), 49.

57. John McCumber, "The Failure of Rational Choice Theory," *New York Times*, June 19, 2011.

58. David Foster Wallace, *This Is Water: Some Thoughts, Delivered on a Significant Occasion, about Living a Compassionate Life* (New York: Little, Brown, 2009), 117.

59. Dupree and O'Malley, "Metagenomics and Biological Ontology," 835, 841.

60. This is an ethological perspective from the Estonian biosemiotician Jakob von Uexküll, *A Foray Into the Worlds of Animals and Humans* (Minneapolis: University of Minnesota Press, 2010).

61. Parikka, *Insect Media*, xxv.

62. Donna Haraway, "The Promises of Monsters: A Regenerative Politics for Inappropriated Others," in *Cultural Studies* (New York: Routledge, 1992), 298.

63. Juengst, "Metagenomic Metaphors," 130.

64. Nerlich and Hellsten, "Beyond the Human Genome," 27–28.

65. Dupree and O'Malley, "Metagenomics and Biological Ontology," 842.

66. Hird, *The Origins of Sociable Life*, 56.

67. Nerlich and Hellsten, "Beyond the Human Genome," 33.

68. Susanne Knudsen, "Communicating Novel and Conventional Scientific Metaphors: A Study of the Development of the Metaphor of the Genetic Code," *Public Understanding of Science* 14 (2005), 388.

69. Napier, *The Age of Immunology*, 65.

70. Michael Hardt and Antonio Negri, *Commonwealth* (Cambridge, Mass.: Belknap Press of Harvard University Press, 2009), 26.

71. Roberto Esposito, *Bios: Biopolitics and Philosophy* (Minneapolis: University of Minnesota Press, 2008), 13.

72. Bruno Latour, *Reassembling the Social: An Introduction to Actor-Network-Theory* (Oxford: Oxford University Press, 2005), 217–18.

73. Committee on Metagenomics, *The New Science of Metagenomics*, 20.

74. Parikka, *Insect Media*, xxv.

75. Ibid., xxv.

76. Grosz, *Chaos, Territory, Art*.

77. Margulis and Sagan, *Acquiring Genomes*, 90.

78. Dupree and O'Malley, "Metagenomics and Biological Ontology," 834–46.

79. Rosi Braidotti, *Transpositions: On Nomadic Ethics* (Cambridge: Polity Press, 2006), 270.

80. Bennett, *Vibrant Matter*, 112–113.

81. By way of example, a larger ethical project based on metagenomics and the HMP can be outlined. Although outside the scope of the current project, a more detailed investigation into work already being done on bioethics and genomics as well as how language-based social contractarian ideas of ethics fall short could prove useful. How could justice and rights factor into these microbial

relationships? Martha Nussbaum offers a particularly fruitful approach from a liberal Rawlsian perspective for the extension of rights to animals in her book *Frontiers of Justice* (Cambridge, Mass.: Harvard University Press, 2006). Donna Haraway's work offers an ethical approach to the world that includes rights for our "messmates"—or all those who share our metaphorical and literal meals—from dogs to bacteria. See especially *When Species Meet* (Minneapolis: University of Minnesota Press, 2008). For an explicit extension to IR, see Rafi Youatt, "Interspecies Relations, International Relations: Rethinking Anthropocentric Politics," *Millennium: Journal of International Studies* 43, no. 1 (2014): 207–23. I believe in Jane Bennett's "careful anthropomorphizing" in order to begin a conversation with a world that acts upon us and one that we are beholden to interact with ethically in a world that does not speak our language, but that we owe something to nonetheless. We have an attachment to the "human estate" (to use William Connolly's term), but also a responsibility to think beyond it as our actions affect so many on the planet. We may know more than we thought about other species, and we are learning more every day. The decoding of whale syntax, the discovery that crocodiles communicate with sounds above the human hearing range, and of the family corvidae's tool use are cases in point.

82. Bennett, *Vibrant Matter*, 112–13.

83. Committee on Metagenomics, *The New Science of Metagenomics*, 32.

84. Margulis and Sagan, *Acquiring Genomes*, 19.

85. Metagenomics as an information technology, and information technology's metaphors more broadly, has helped us, following Latour in *Reassembling the Social*, "to realize the work going on in actor-making. It's now much easier to not consider the actor as a subject endowed with some primeval interiority, which turns its gaze toward an objective world made of brute things to which it should resist or out of which it should be able to cook up some symbolic brew. Rather, we should be able to observe empirically how an anonymous and generic body is made to be a person: the more intense the shower of offers of subjectivities, the more interiority you get" (208).

3. States in Nature, Nature in States

1. William Parker and Jeff Ollerton, "Evolutionary Biology and Anthropology Suggest Biome Reconstitution as a Necessary Approach Toward Dealing with Immune Disorders," *Evolution, Medicine, and Public Health*, 1 (2013): 89–103.

2. William Parker, "Reconstituting the Depleted Biome to Prevent Immune Disorders," accessed September 25, 2016, http://evmedreview.com/?p=457.

3. Caroline Hadley, "Should Auld Acquaintance Be Forgot?," *European Molecular Biology Association* 5, no. 12 (2004), 1123.

4. See Bruno Latour, *The Pasteurization of France* (Cambridge, Mass.: Harvard University Press, 1988).

5. The Institute of Medicine, *Ending the War Metaphor: The Changing Agenda for Unraveling the Host-Microbe Relationship* (Washington, D.C.: National Academies Press, 2006), 14.

6. Ibid.

7. A. David Napier, *The Age of Immunology: Conceiving a Future in an Alienating World* (Chicago: University of Chicago Press, 2003).

8. *Review of International Studies* 30, no. 2 (April 2004).

9. Patrick Thaddeus Jackson, "Forum Introduction: Is the State a Person? Why Should We Care?," *Review of International Studies* 30, no. 2 (April 2004): 257.

10. Bruno Latour, *Reassembling the Social: An Introduction to Actor-Network-Theory* (Oxford: Oxford University Press, 2005), 71.

11. Ibid., 79.

12. Joost Van Loon, *Risk and Technological Culture: Towards a Society of Virulence* (London: Routledge, 2002), 51.

13. Anthony Burke, *Beyond Security, Ethics and Violence: The War against the Other* (New York: Routledge, 2007), 39.

14. Van Loon, *Risk and Technological Culture*, 246.

15. R. B. J. Walker, *After the Globe, Before the World* (London: Routledge, 2010), 185. See also R. B. J. Walker, *Inside/Outside: International Relations as Political Theory* (Cambridge: Cambridge University Press, 1993).

16. Committee on Metagenomics, *The New Science of Metagenomics: Revealing the Secrets of Our Microbial Planet* (Washington, D.C.: National Academies Press, 2007), 22.

17. Anne Maczulak, *Allies and Enemies: How the World Depends on Bacteria* (Upper Saddle River, N.J.: FT Press, 2010), 15–16.

18. Walker, *Inside/Outside*, 62, 64.

19. Ibid., 65.

20. Alexander Hinton, "The Dark Side of Modernity: Toward an Anthropology of Genocide," in *Annihilating Difference: The Anthropology of Difference*, ed. Alexander Hinton (Berkeley: University of California Press, 2002), 1.

21. Laurence Whitehead, "Enlivening the Concept of Democratization: The Biological Metaphor," *Perspectives on Politics* 9, no. 2 (2011): 295.

22. Burke, *Beyond Security, Ethics and Violence*, 39.

23. A. David Napier, *The Age of Immunology: Conceiving a Future in an Alienating World* (Chicago: University of Chicago Press, 2003), 144.

24. Whitehead, "Enlivening the Concept of Democratization," 295.

25. Hinton, "The Dark Side of Modernity," 1.

26. Alfred I. Tauber and Leon Chernyak, *Metchnikoff and the Origins of Immunology: From Metaphor to Theory* (New York: Oxford University Press, 1991), xv.

27. In law, immunity can be sovereign, parliamentary, diplomatic, prosecutorial, or judicial. Sovereign immunity has become a debated concept in international relations. Claims of the eroding legitimacy of sovereign immunity and subsequent decisions to prosecute heads of state for war crimes and crimes against humanity coupled with the creation of the International Criminal Court has further supported the idea that individuals, regardless of their status, should be accountable and personally liable for judgment and prosecution. For this book, the history of the term's crossing the boundary between medical science and law serves as a further example of the reciprocal relationship between the two discourses.

28. Tauber and Chernyak, *Metchnikoff and the Origins of Immunology*, xiv.

29. Ibid., xvi.

30. Elling Ulvestad, *Defending Life: The Nature of Host-Parasite Relations* (Dordrecht: Springer, 2007), 15.

31. Donna Haraway, "The Promises of Monsters: A Regenerative Politics for Inappropriated Others," in *Cultural Studies* (New York: Routledge, 1992), 321.

32. The Institute of Medicine, *Ending the War Metaphor*, 3–4.

33. Centers for Disease Control, "Detection of Enterobacteraceae Isolates Carrying Metallo-Beta Lactamase," accessed September 25, 2016, http://www.cdc.gov/mmwr/preview/mmwrhtml/mm5924a5.htm.

34. Richard Knox, "Super-Resistant Gonorrhea Strain Found in Japan," accessed September 25, 2016, http://www.npr.org/blogs/health/2011/07/12/137797153/super-resistant-gonorrhea-strain-found-in-japan.

35. The Institute of Medicine, *Ending the War Metaphor*, 4.

36. See also Brigitte Nerlich and Richard James, "'The Post-Antibiotic Apocalypse' and the 'War on Superbugs': Catastrophe in Microbiology, Its Rhetorical Form and Political Function," *Public Understanding of Science* 18, no. 5 (2009): 574–90; and for public perceptions of emerging infectious diseases, see *Public Understandings of Science* (July 2011).

37. The Institute of Medicine, *Ending the War Metaphor*, 2.

38. Ulvestad, *Defending Life*, xv.

39. Alfred I. Tauber, "The Immune System and Its Ecology," *Philosophy of Science* 75 (2008): 232.

40. Whitehead, "Enlivening the Concept of Democratization," 296.

41. The Institute of Medicine, *Ending the War Metaphor*, 2.

42. Ecological views, however, remain contentious, even with an increase in functional understandings of the immune system.

43. The Institute of Medicine, *Ending the War Metaphor*, 2.

44. Ulvestad, *Defending Life*, 15.

45. The Institute of Medicine, *Ending the War Metaphor*, 4.

46. Ibid., 1.
47. Ulvestad, *Defending Life*, 15–16.
48. Hadley, "Should Auld Acquaintance Be Forgot?," 1123.
49. G. A. W. Rook, "Review Series on Helminth, Immune Modulation and the Hygiene Hypothesis: The Broader Implications of the Hygiene Hypothesis," *Immunology* 126 (2008), 8.
50. Parker, "Reconstituting the Depleted Biome."
51. Ibid.
52. Rook, "Review Series on Helminth," 8.
53. Parker, "Reconstituting the Depleted Biome."
54. Hadley, "Should Auld Acquaintance Be Forgot?," 1123.
55. Ibid., 1124.
56. Eric Niller, "Got Allergies? Take a Worm," accessed September 25, 2016, http://news.discovery.com/human/parasites leeches-maggots-worms-medicine .html.
57. Rob Stein, "Immune Systems Increasingly on Attack," *Washington Post*, March 4, 2008.
58. It was originally thought that man picked up helminths from domesticated animals, but it has since been put forth, due to doubts about the short amount of time in evolutionary terms (about five hundred generations) in which man has lived with animals, that man may have picked up helminths from other carrion eaters around one million years ago. Man then transferred these helminths to domesticated animals.
59. G. A. W. Rook, "The Hygiene Hypothesis and the Increasing Prevalence of Chronic Disease Disorders," *Transactions of the Royal Society of Tropical Medicine and Hygiene* 101, no. 1 (2007): 1072–74.
60. Niller, "Got Allergies?"
61. Stein, "Immune Systems Increasingly on Attack."
62. The use of the word *person* brings in another level to the discussion that this book cannot fully address. Briefly, the idea of a person has a legal and social meaning beyond just another way to say individual or human being. A "person" is recognized as a subject under the law. In part, a state is classified as a person in order for it to be accorded a certain legal status that affords it greater legal protections and rights; the state is a metaphysical person. In the United States, this is accomplished through the Fourteenth Amendment to the Constitution. Many political and moral implications follow from this legal classification both domestically and internationally.
63. Haraway, "The Promises of Monsters," 323.
64. Parker, "Reconstituting the Depleted Biome."
65. Hadley, "Should Auld Acquaintance Be Forgot?," 1122.

66. Hillis Miller, "The Critic as Host," *Critical Inquiry* 3, no. 3 (1977), 442.

67. Michel Serres, *The Parasite* (Baltimore: Johns Hopkins University Press, 1982), 7.

68. Miller, "The Critic as Host," 442.

69. Serres, *The Parasite*, 34.

70. Sharon Shattuck, *Parasites: A User's Guide* (2010), DVD. For more information, see, http://parasites-film.com/.

71. Van Loon, *Risk and Technological Culture*, 251.

72. Serres, *The Parasite*, 12.

73. David F. Bell, "Untitled Review of Michel Serres, *Le Parasite*," *MLN* 96, no. 4 (1981): 886.

74. Serres, *The Parasite*, 33.

75. Bell, "Untitled Review of Michel Serres," 886.

76. This idea was proposed by P. P. Grime, "Competitive Exclusion in Herbaceous Vegetation," *Nature* 242 (1973): 344–47; and again by J. H. Connell, "Diversity in Tropical Rain Forests and Coral Reefs," *Science* 199 (1978): 1302–10. Debate over this theory is ongoing. See Stephen Roxburgh, Katrina Shea, and J. Bastow, "The Intermediate Disturbance Hypothesis: Patch Dynamics and Mechanisms of Species Coexistence," *Ecology* 85, no. 2 (2004): 359–71.

77. Whitehead, "Enlivening the Concept of Democratization," 274.

78. Joost Van Loon, "Parasite Politics: On the Significance of Symbiosis in Theorizing Community Formations," in *Politics at the Edge*, ed. Chris Pierson and Simon Tormey (London: Palgrave McMillan, 2000), 250.

79. Must notably, Huntington and his "Clash of Civilizations" theory argues that difference is the cause of conflict, and the homogeneous sameness of culture promotes more peaceful relations within the culture. Difference is not seen as "stimulating" or necessary, but as a destabilizing variable in cultural relations. See Samuel P. Huntington, *The Clash of Civilizations and the Remaking of World Order* (New York: Simon & Schuster, 1996).

80. Author unknown, "The Hygiene Hypothesis or Old Friends Hypothesis," accessed September 25, 2016, http://www.hygienehypothesis.com/.

81. Ibid.

82. Shattuck, *Parasites: A User's Guide*.

83. Van Loon, "Parasite Politics," 242.

84. Keith Ansell Pearson, *Viroid Life: Perspectives on Nietzsche and the Transhuman Condition* (London: Routledge, 1997), 124.

85. Tauber, "The Immune System and Its Ecology," 231.

86. Ibid., 232.

87. Van Loon, *Risk and Technological Culture*, 77–78.

88. Roberto Esposito, *Bios: Biopolitics and Philosophy* (Minneapolis: University of Minnesota Press, 2008), 18.

89. Whitehead, "Enlivening the Concept of Democratization," 294.

90. George Lakoff, "Metaphor and War: The Metaphor System Used to Justify War in the Gulf," *Viet Nam Generation Journal* 3 (1991), accessed February 2, 2017, http://www2.iath.virginia.edu/sixties/HTML_docs/Texts/Scholarly/Lakoff_Gulf_Metaphor_1.html 9.

91. Whitehead, "Enlivening the Concept of Democratization," 294.

4. Posthuman Politics

1. Steven Bernstein, Richard Ned Lebow, Janice Gross Stein, and Steven Weber, "God Gave Physics the Easy Problems: Adapting Social Science to an Unpredictable World," *European Journal of International Relations* 6, no. 1 (2000): 44.

2. Diana Coole and Samantha Frost, eds., *New Materialisms: Ontology, Agency, and Politics* (Durham, N.C.: Duke University Press, 2010), 11.

3. Bernstein et al., "God Gave Physics the Easy Problems," 48.

4. Erika Cudworth and Stephen Hobden, *Posthuman International Relations* (London: Zed Books, 2011).

5. Laurence Whitehead, "Enlivening the Concept of Democratization: The Biological Metaphor," *Perspectives on Politics* 9, no. 2 (2011), 296.

6. Ibid., 295.

7. Michel Foucault, *Society Must Be Defended: Lectures at the College de France, 1975–1976* (New York: Picador, 2003), 168.

8. Whitehead, "Enlivening the Concept of Democratization," 295.

9. S. Eben Kirksey and Stefen Helmreich, "The Emergence of a Multispecies Ethnography," *Cultural Anthropology* 25, no. 4 (2010): 545–76.

10. Irun R. Cohen and David Harel, "Explaining a Complex Living System: Dynamics, Multi-Scaling and Emergence," *Journal of the Royal Society Interface* 4, no. 13 (2007): 176.

11. Cary Wolfe, *What Is Posthumanism?* (Minneapolis: University of Minnesota Press, 2010), xv.

12. Ibid., xvi.

13. Ibid., xi.

14. N. Katherine Hayles, *How We Became Posthuman: Virtual Bodies in Cybernetics, Literature, and Informatics* (Chicago: University of Chicago Press, 1999), 285.

15. Ed Cohen, "Immune Communities, Common Immunities," *Social Text* 26, no. 1 (Spring 2008): 99.

16. Foucault, *Society Must Be Defended*.

17. See Zakiyyah Iman Jackson, "New Directions in the Theorization of Race and Posthumanism," *Feminist Studies* 39, no. 3 (2013): 669–85; Mel Y. Chen, *Animacies: Biopolitics, Racial Mattering, and Queer Affect* (Durham, N.C.: Duke University Press, 2012); Kalpana Rahita Seshadri, *HumAnimal: Race, Law, and*

Language (Minneapolis: University of Minnesota Press, 2012); and Michael Lunblad, *The Birth of a Jungle: Animality in Progressive-Era U.S. Literature* (Oxford: Oxford University Press, 2013).

18. Rosi Braidotti, *The Posthuman* (Cambridge: Polity Press, 2013), 1.

19. Ibid.; Wolfe, *What Is Posthumanism?*

20. Alexander Weheliye, *Racializing Assemblages, Biopolitics, and Black Feminist Theories of the Human* (Durham, N.C.: Duke University Press, 2014).

21. James C. Scott, *Seeing Like a State: How Certain Schemes to Improve the Human Condition Have Failed* (New Haven, Conn.: Yale University Press, 1998).

22. Siba N. Grovogui, *Beyond Eurocentrism and Anarchy: Memories of International Order and Institutions* (New York: Palgrave Macmillan, 2006).

23. Braidotti, *The Posthuman*, 2.

24. Andrew Pickering and Keith Guzik, *The Mangle in Practice* (Durham, N.C.: Duke University Press, 2008), 4.

25. Committee on Metagenomics, *The New Science of Metagenomics: Revealing the Secrets of Our Microbial Planet* (Washington, D.C.: National Academies Press, 2007), 12.

26. Bruno Latour, "Whose Cosmos, Which Cosmopolitics? Comments on the Peace Terms of Ulrich Beck," *Common Knowledge* 10, no. 3 (2004): 450–62.

27. A. P. Alivisatos et al., "A Unified Initiative to Harness Earth's Microbiomes," *Science* 350, no. 6250 (October 30, 2015).

28. Nicole Dubilier, Margaret McFall-Ngai, and Liping Zhao, "Microbiology: Create a Global Microbiome Effort," *Nature* 526 (October 29, 2015).

29. Ibid.

30. Michel Foucault, *The History of Sexuality, Volume 1: An Introduction* (New York: Vintage Books, 1990).

31. Zygmunt Bauman, "The Uniqueness and Normality of the Holocaust," in *Modernity and the Holocaust* (Ithaca, N.Y.: Cornell University Press, 1989).

32. Roberto Esposito, *Bios: Biopolitics and Philosophy* (Minneapolis: University of Minnesota Press, 2008), 54–55.

33. Alfred I. Tauber, "The Immune System and Its Ecology," *Philosophy of Science* 75 (2008): 229.

34. Ed Cohen, *A Body Worth Defending: Immunity, Biopolitics and the Apotheosis of the Modern Body* (Durham, N.C.: Duke University Press, 2009), 3.

35. Tauber, "The Immune System and Its Ecology," 234.

36. Ibid.

37. Alfred I. Tauber and Leon Chernyak, *Metchnikoff and the Origins of Immunology: From Metaphor to Theory* (New York: Oxford University Press, 1991), xiv.

38. Cohen, "Immune Communities, Common Immunities," 110.

39. Ibid.

40. Cohen and Harel, "Explaining a Complex Living System," 177.

41. Ibid.
42. Ibid., 181.
43. Ibid., 175.
44. Sean R. Eddy, "'Antedisciplinary' Science," *PLoS Computational Biology* 1, no. 1 (June 2005), accessed February 2, 2017, http://dx.doi.org/10.1371/journal.pcbi.0010006.
45. Ibid.

Coda

1. See especially Erika Cudworth and Steohen Hobden, "Of Parts and Wholes: International Relations beyond the Human," *Millennium: Journal of International Studies* 41, no. 3 (2013): 430–50.

2. While the reader might expect a footnote on the "classics" in IR theory here, I refrain and say that these texts have their rhetorical power, too. Much like a zombie, one bite can infect an otherwise free-thinking person. These texts often feel like "truth," and this is exactly what I ask the reader to question. How do these texts presume to explain the world? How are they assumed to be true?

3. Anthony Burke, Simon Dalby, Stefanie Fishel, Danile Levine, and Audra Mitchell, "Planet Politics: A Manifesto from the End of IR," *Millennium: Journal of International Studies* 44, no. 3 (2016): 499–523. See also the 2015 Millennium Conference Keynote by Bruno Latour, "Onus Orbis Terrarum: About a Possible Shift in the Definition of Sovereignty," *Millennium: Journal of International Studies* 44, no. 3 (2016): 305–20. Also Daniel Levine, *Recovering International Relations: The Promise of Sustainable Critique* (Oxford: Oxford University Press, 2012), frames international relations as a practice and a vocation and traces productive shifts in IR's understanding of itself.

4. Michael Hardt and Antonio Negri, *Commonwealth* (Cambridge, Mass.: Belknap Press of Harvard University Press, 2009), 71.

5. Nigel Thrift, "Halos: Making More Room in the World for New Political Orders," in *Political Matter*, ed. Bruce Braun and Sarah J. Whatmore (Minneapolis: University of Minnesota Press, 2010), 139.

6. Gilles Deleuze and Felix Guattari use the phrase "nomad science" or "minor sciences" to contrast with the "royal sciences" or those sciences that uphold the state form of organization. See Deleuze and Guattari, *A Thousand Plateaus: Capitalism and Schizophrenia* (Minneapolis: University of Minnesota Press, 1987).

7. Walt Whitman, "Kosmos," in *Leaves of Grass* (Philadelphia: David McKay, 1891–92), 303–4.

Index

Actant, 44–46, 53, 61, 71–72, 75, 79, 86, 96; state as, 78–80, 114
Actor-Network-Theory (ANT), 78
Anthropocene, 1, 21, 116
Anthropocentrism, 6
Antibiotic resistant bacteria, 85, 87
Arendt, Hannah, 20, 22, 39
Assemblage, 39, 42, 47, 56, 104; bodies as, 6, 38–39, 67, 68, 99, 113; model for IR, 21, 72–73
Autopoiesis, 100

Balance of power, 48; as metaphor, 48–49, 51
Bare life, 5, 104
Bennett, Jane, 6, 38, 47
Biome depletion theory, 76, 88–91; application to the state, 81, 83, 90, 96; parasites and, 91–96
Biopolitics, 21–23; ecological immunity and, 108–9; immunity and, 106, 107–8
Biopower, 21, 108
Bodies politic, 23, 43, 61, 72
Body, 2, 41, 62; affect and, 38; becoming, 14, 38–39, 42, 63–64, 74; changing perceptions of, 71, 73, 75, 102–3; container metaphor, 51, 62–63; immunity and, 83, 84, 90, 91, 106–9; IR and, 10, 25–27, 48, 72, 114; lively vessel, 57, 62; materiality and, 5, 16, 25, 27, 40–41; metagenomics, 58–60, 71; parasite and, 92–93, 94; superorganism and, 64, 70
Body politic, 20, 23, 62, 71, 98, 102, 113, 116; ecological immunity, 91, 94, 96; immunity and, 40; IR and, 26, 28–33, 113; metagenomics and, 57; metaphor, 2–3, 4, 11–12, 18–19, 39, 47, 115; the state and, 76, 80, 91
Braidotti, Rosi, 21, 103–4
Brain-in-a-vat, 35, 40, 47.
Burke, Anthony, 31, 80

Canguilhelm, Georges, 51
Capitalism, 11, 40, 69, 115
Coevolution, 7, 23, 52, 65, 75, 77, 92
Cohen, Ed, 8, 15, 107, 109
Committee on Metagenomics, 60, 72, 81, 105
Connolly, William, 12, 37, 42, 132n81
Constructivism, 8, 36, 41, 79
Contaminated state, 15, 75, 91–97, 98, 113, 115; health and, 76, 90; metaphor and, 15, 95; posthuman and, 102–3

Deleuze, Gilles, 34, 64, 139n6
Democracy, 4, 10, 19, 25

Depleted biome. *See* Biome depletion theory
Descartes, René, 37. *See also* Dualism
Deutsch, Karl, 48–49
DNA, 14, 65, 99
Dualism (Cartesian), 14, 23, 27, 35, 37, 40, 104

Entanglement, 19, 98; ethics and, 21, 71, 94; metaphor and, 46–47
Esposito, Roberto, 40, 107
Ethics, 14, 22, 39, 74–75, 125n35; entanglement and, 21, 71, 94; posthumanism and, 94; Spinozan, 39; state and, 82–83
Eukaryotes, 65, 93. *See also* Helminths
Evolution, 8, 23, 55, 58, 60, 100; Darwinian, 8, 83–84, 95, 101; microbes and, 77, 87–88, 89–90; symbiosis and, 64–66, 75. *See also* Symbiogenesis

Feyerabend, Paul, 15, 36–37
Forum on Microbial Threats, 85, 86–88; metaphor and, 84–87
Foucault, Michel, 26, 34, 40, 106

Geneva Conventions, 26

Haraway, Donna, 6, 23, 37, 69, 84, 91, 132n81
Helminths, 77, 88, 89, 90, 95, 135n58. *See also* Eukaryotes
Hird, Myra, 71
Hobbes, Thomas, 26, 32, 37, 48, 107; body politic and, 28–31
Homo contaminatus, 75
Human Genome Project (HGP), 70

Humanitarian intervention, 49–50
Human Microbiome Project (HMP), 57, 58–61, 70–71, 76, 91

Immunity, 4; biopolitics and, 106–9; contaminated state and, 90, 95–97, 100; ethics and, 93, 109; history of, 83–85; law and, 27, 83, 108, 134n27; Methnikoff and, 83–86; organisms and, 84; parasites and, theories of, 77–78, 83. *See also* Immunology
Immunology, 43, 83, 100, 107; ecological theories of, 85–87, 90, 91, 92, 93–94, 108; pure culture and, 82–83; state and, 95–97; war metaphor and, 84–85, 87. *See also* Immunity
Intermediate-disturbance hypothesis, 94
International Relations (IR), 1, 37, 98, 113; biological metaphor and, 100, 109–11; bodies and, 2, 25–27, 38–39; body politic and, 26, 72; critical approaches and, 3–5, 33–34, 36–38, 114; disciplinary failure and, 4; ethics, 74–75, 94; materialism and, 42, 78; metaphor, 28–29, 47–49, 51–55, 61, 97, 99–106, 114; scientific approaches to, 4–6, 11–15, 36, 42–43, 99, 116; State as person and, 57, 78–81, 88, 114–15; state centrism and, 8–11, 27–28, 33; STS and, 43–46; theory and, 16, 33–34

Jackson, Patrick Thaddeus, 9, 78
Johnson, Mark, 17, 51

Lakoff, George, 17, 18, 49, 50, 51, 76
Latour, Bruno, 13, 21, 22, 35, 44, 72, 110, 132n85; actants and, 45, 53, 78–79, 114; Thomas Hobbes and, 28–29

Leviathan, 28–31, 32
Little, Richard, 14, 35
Lively vessel, 57, 61–68, 75, 101, 104, 113; metaphor and, 15, 115; posthuman and, 102–3; state and, 62, 98, 115
Locke, John, 31, 37, 49
Love, 6, 17, 18, 20, 37

Margulis, Lynn, 23, 63, 64, 65
Marks, Michael, 52
Materialism, 15–16, 28, 42–43, 47, 61, 73, 78
Metagenomics, 13, 43, 56–58, 65, 85, 105, 132n85; immunology and, 96, 100; metaphor and, 61, 68–72, 115; symbiosis and, 67–68
Metaphorogenesis, 46, 51–54
Metaphors: agent-structure, 114; balance of power and, 48–49, 51; biological, 98, 99–101; cognitive, 17–18, 47, 54; container, 48, 51, 62–64, 75, 105; contaminated state and, 15, 95; corporeal, 3, 16, 23, 25, 29; Forum on Microbial Threats and, 84–87; International Relations and, 28–29, 47–49, 51–55, 61, 97, 99–106, 114; lively vessel and, 15, 115; ontology and, 39, 67, 71, 96; parasites and, 92–96; science and, 17; state and, 5, 40, 50–52, 71, 110; synecdoche, 18, 28; transformative, 16, 18, 51–55, 115; war as, 49–50, 52, 77, 92
Metchnikoff, Elie, 83–85, 107
Microbe, 2, 5–6, 47, 55–56, 65, 72–73; immune system and, 77, 86, 87, 89, 96; metagenomics and, 13, 57–58, 81, 105; metaphor and, 48, 57, 59, 60–61, 71, 76; changing perceptions of, 60–61, 70–71

Microbiome, 58, 90, 113; body politic and, 57; global diversity of, 105–6; human body and, 5, 43, 59–60, 103; metaphor and, 68–75, 106; self-identity and, 60, 64, 115
Morton, Timothy, 14, 35

Nancy, Jean-Luc, 26–27
Nation-state, 33, 80, 82, 107
Nested sets of permeable bodies, 43, 64, 76, 98
Nietzsche, Friedrich, 19, 116
Noise, 92, 93–94. *See also* Parasites

Old friends hypothesis, 76–77, 88, 91; parasites and, 92–94. *See also* Biome depletion theory
Ontology, 8, 15, 58, 104; metaphor and, 39, 67, 71, 96; state and, 28, 79
Organism, 31, 32, 55, 65, 73, 108; body and the, 14, 41, 62–63; immunity and, 83–85, 87–88, 91–92, 108; individuality and, 68–70, 76; metagenomics and, 57–58, 66, 75, 77–78; metaphor and, 48, 60, 67, 68, 69–70, 107; symbiogenesis, 65–66; systems biology and, 109–10

Parasites, 5, 21, 66, 77, 92–96, 115; biome depletion theory and, 89–90, 91–95; contaminated states and, 96–97; ethics and, 94–95; immigration and, 95–96; metaphor and, 95–97; noise and, 93–94; subject creation and, 93
Peace of Westphalia, 32, 35
Physics, 15, 98, 99, 100
Pickering, Andrew, 7, 105
Pluralism, 20, 22, 37, 95

Posthuman, 7–8, 63, 97, 99, 101–5; animal, 103, 104; becoming and, 101–6; body and, 103, 109; microbiome, 99, 105–6; race, 101–2, 103–4
Pure culture paradigm, 81–82

Realism, 33, 79, 123n7
Rousseau, Jean-Jacques, 31

Sagan, Carl, 98
Sagan, Dorion, 23, 63, 64, 65
Science, technology, and society studies (STS), 43–44. *See also* Science studies
Science studies, 47, 78; IR and, 43–46
Serres, Michel, 92–94
Sovereignty, 9–10, 32, 80–81, 104, 107; body politic and, 29, 31
Spinoza, Benedict de, 21, 37, 38, 39, 45
State, 4, 23, 27, 31–32; aseptic, 81–83, 90, 98; contaminated, 91, 95–97; critique of, 15, 25, 35, 36, 44, 61; culturally produced, 72; depleted biome and, 96–97; health, 14, 76–77; IR and, 1, 8–11, 27–28, 33; metaphor and, 5, 40, 50–52, 71, 110; ontology and, 28, 79; personhood and, 26, 57, 78–81, 88, 114–15, 135n62
Symbiogenesis, 65, 73, 77
Symbiopoiesis, 73–76
Symbiosis, 65, 67; parasites and, 95–97
Systems biology, 101–5; metaphor and, 99–100, 110–11

Translations, 21, 44, 46, 79–80, 114

Unified Microbiome Initiative Consortium, 105–6

Walker, R. B. J., 10, 52–53, 81
Wendt, Alexander, 78, 114
Whitehead, Laurence, 86; biological metaphors and, 100–101
Whitman, Walt, 1, 19–21, 98, 116–17; *Leaves of Grass*, 1, 19, 25, 55, 76, 98, 113

Zoë, 5. *See also* Bare life

STEFANIE R. FISHEL is assistant professor of gender and race studies at the University of Alabama.

Made in the USA
Lexington, KY
30 July 2017